Prominent Properties of Composites

Prominent Properties of Composites

Edited by **Gerald Brooks**

New York

Published by NY Research Press,
23 West, 55th Street, Suite 816,
New York, NY 10019, USA
www.nyresearchpress.com

Prominent Properties of Composites
Edited by Gerald Brooks

International Standard Book Number: 978-1-63238-378-5 (Hardback)

Contents

Preface

This book has been an outcome of determined endeavour from a group of educationists in the field. The primary objective was to involve a broad spectrum of professionals from diverse cultural background involved in the field for developing new researches. The book not only targets students but also scholars pursuing higher research for further enhancement of the theoretical and practical applications of the subject.

Composites are a class of materials which possess some unique and extraordinary properties. They have received a lot of attention recently due to their role in the field of material research and also because of the advent of new forms of composites such as nanocomposites and bio-medical composites. Composite materials can be utilized in various industries such as aerospace and construction industries. This book primarily deals with the production and property classification of various composites. It covers nanocomposites, damages and fractures-theoretical and numerical modelling, and composites - physical features, designing, processing and manufacturing methods.

It was an honour to edit such a profound book and also a challenging task to compile and examine all the relevant data for accuracy and originality. I wish to acknowledge the efforts of the contributors for submitting such brilliant and diverse chapters in the field and for endlessly working for the completion of the book. Last, but not the least; I thank my family for being a constant source of support in all my research endeavours.

Editor

Nanocomposites

Graphene Nanocomposites

Mingchao Wang, Cheng Yan and Lin Ma

Additional information is available at the end of the chapter

1. Introduction

Graphene, one of the allotropes (diamond, carbon nanotube, and fullerene) of carbon, is a monolayer of honeycomb lattice of carbon atomsdiscovered in 2004. The Nobel Prize in Physics 2010 was awarded to Andre Geim and Konstantin Novoselov for their ground breaking experiments on the two-dimensional graphene [1]. Since its discovery, the research communities have shown a lot of interest in this novel material owing to its unique properties. As shown in Figure 1, the number of publications on graphene has dramatically increased in recent years. It has been confirmed that graphene possesses very peculiar electrical properties such as anomalous quantum hall effect, and high electron mobility at room temperature (250000 cm^2/Vs). Graphene is also one of the stiffest (modulus ~1 TPa) and strongest (strength ~100 GPa) materials. In addition, it has exceptional thermal conductivity (5000 Wm^{-1}K^{-1}). Based on these exceptional properties, graphene has found its applications in various fields such as field effect devices, sensors, electrodes, solar cells, energy storage devices and nanocomposites. Only adding 1 volume per cent graphene into polymer (e.g. polystyrene), the nanocomposite has a conductivity of ~0.1 Sm^{-1}[2], sufficient for many electrical applications. Significant improvement in strength, fracture toughness and fatigue strength has also been achieved in these nanocomposites [3-5]. Therefore, graphene-polymer nanocomposites have demonstrated a great potential to serve as next generation functional or structural materials.

Relatively, limited research has been conducted to understand the intrinsic structure-property relationship in graphene based composites such as graphene-polymer nanocomposites. The mechanical property enhancement observed in graphene-polymer nanocomposites is generally attributed to the high specific surface area, excellent mechanical properties of graphene, and its capacity to deflect crack growth in a far more effectively way than one-dimensional (e.g. nanotube) and zero-dimensional (e.g. nanoparticle) fillers [5]. On the other hand, the graphene sheets or thin platelets dispersed in polymer matrix may create wavy or wrinkled structures that tend to unfold rather than stretch under applied loading.

This may severely reduce their stiffness due to weak adhesion at the graphene-polymer interfaces [6]. However, a wrinkled surface texture could create mechanical interlocking and load transfer between graphene and polymer matrix, leading to improved mechanical strength [7]. Furthermore, structural defects and stability of graphene can significantly influence the graphene-polymer interfacial behaviour. Therefore, further work is required to understand the structure-property relationship in graphene and the graphene-polymer interface behaviour.

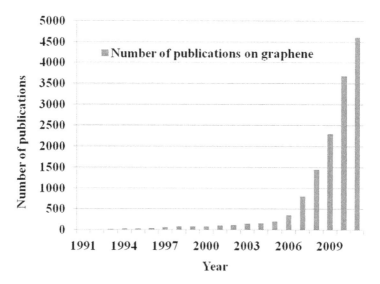

Figure 1. Number of publications on graphene in past 20 years.

2. Graphene

2.1. Mechanical and electrical properties

Mechanical properties

Graphene, a special monolayer of hexagon-lattice, are even stiffer and stronger than carbon nanotube (CNT). By nanoindentation on a free-standing monolayer graphene, the Young's modulus and intrinsic strength were estimated as $\sim$1.0 TPa and σ_{int}=130 GPa at a strain of ε_{int}=0.25 [8,9]. Atomistic simulations demonstrated size and chirality dependent elastic properties in graphene nanoribbons [10,11]. The size effect on Young's modulus is negligible when the diagonal length of a graphene nanoribbon is over 10.0 nm. The maximum Cauchy (true) stress and fracture strain for graphene loaded in the armchair direction were estimated as 102 GPa and 0.13, respectively. Higher values were observed in the zigzag direction, i.e., 129 GPa and 0.20, respectively. Besides size and chirality dependence, temperature also shows significant influence on the mechanical properties of graphene. Zhao et al. suggested[12] that Young's modulus does not vary significantly with temperature until about 1200 K, beyond which graphene becomes softer. The fracture strength and fracture strain decrease significantly with the increase with temperature [12]. Even though monolayer graphene is generally regarded as an ideal structure for practical applications, graphene flakes with few layers are often present in the routine of synthesis, such as mechanical exfoliation. It has been confirmed the layer number is another noticeable factor in dictating the mechanical properties. Table 1 summarizes the intrinsic mechanical properties of the single, bilayer and trilayer graphene.

Method	Material	Mechanical properties	References
AFM	Monolayer graphene	$E = 1 \pm 0.1$ TPa	[8]
		$\sigma_{int} = 130 \pm 10$ GPa at $\varepsilon_{int} = 0.25$	
Raman	Graphene	Strain ~1.3% in tension	[13]
		Strain ~0.7% in compression	
AFM	Monolayer	E= 1.02 TPa; σ = 130 GPa	[14]
	Bilayer	E= 1.04 TPa; σ = 126 GPa	
	Trilayer graphene	E= 0.98 TPa; σ = 101 GPa	

Table 1. Mechanical properties of graphene.

In a single graphene sheet, the sp^2 hybridized carbon atoms are arranged in hexagonal fashion. A single hexagonal ring comprises of three strong in-plane sigma bonds p_z orbitals perpendicular to the planes. Different graphene layers are bonded by weak p_z interaction. As a result, the hexagonal structure is generallystable but delamination can occur between the graphene layers when subjected to shear stresses. For example, scotch tape was used to obtain single graphene sheet by peeling bulk graphite layer by layer [1]. In general, the interaction between graphene and other material is considered to be in the form of non-bonded van der Waals attraction. For example, the graphene-SiO_2 adhesion energy estimated by pressurized blister tests is about 0.45 ± 0.02 Jm^{-2} for monolayer graphene and 0.31 ± 0.03 Jm^{-2} for two- to five-layer graphene sheets [15]. These values are greater than the adhesion energies measured in typical micromechanical structures and are comparable to solid-liquid adhesion energies. This can be attributed to the extreme flexibility of graphene, which allows it to conform to the topography of even the smoothest substrates, thus making its interaction with the substrate more liquid-like rather than solid-like.

Electrical transport properties

As a semiconductor with zero band gap, graphene has unusual charge carriers that behave as massless relativistic particles (Dirac fermions), which is different from electrons when subjected to magnetic fields and has the anomalous integer quantum Hall Effect (QHE) [16]. This effect was even observed at room temperature [17]. The band structure of single layer graphene exhibits two bands which intersect at two in equivalent point K and K_0 in the reciprocal space. Near these points electronic dispersion resembles that of the relativistic Dirac electrons. K and K_0 are referred as Dirac points where valence and conduction bands are degenerated, making graphene a zero band gap semiconductor.

Another important characteristic of single-layer graphene is its ambipolar electric field effect at room temperature, which is charge carriers can be tuned between electrons and holes by applying a required gate voltage [1,18]. In positive gate bias the Fermi level rises above the Dirac point, which promotes electrons populating into conduction band, whereas in negative gate bias the Fermi level drops below the Dirac point promoting the holes in valence band.

2.2. Structural defect

Recently, different synthesis methods have been developed to produce high quality graphene such as chemical vapor-deposition (CVD)[19-21] and epitaxial growth[22,23] on metal or SiC substrates. However, various defects and impurities are often introduced into graphene during the processing. The second law of thermodynamics also indicates the presence of a certain amount of disorder in crystalline materials. Like other crystalline materials, it is expected the defects and impurities may strongly influence the electrical, mechanical and thermal properties of graphene. Structural defects, such as Stone-Wales (S-W) defect and vacancies in graphene, can significantly reduce its intrinsic strength. Quantized fracture mechanics (QFM) as well as molecular dynamics (MD) simulations demonstrated that even one vacancy can lead to strength loss by 20% of pristine graphene [11]. Zheng et al. [24] found that Young's modulus depends largely on the degree of functionalization and molecular structure of the functional groups attached to a graphene sheet, attributed to the binding energy between the functional groups and the graphene, as well as sp^2-to-sp^3 bond transition. This was also confirmed in the graphene with hydrogen function groups [25].

On the other hand, imperfection in graphene can be used to tailor the properties of graphene and achieve new functions[26,27]. Defects in graphene are divided into two different types, namely *intrinsic* and *extrinsic*. Imperfection without the presence of foreign atoms is referred to as intrinsic type, and other is referred to as extrinsic type. In terms of dimensionality, defects in graphene can also be categorized as point defect (0D) and line defect (1D). In this section, we will review the formation of several typical *intrinsic* lattice defects in graphene.

Point defects

One of the unique properties of graphene is its ability to reconstruct the atom arrangement by forming non-hexagonal rings. The simplest example is SW defect [28], which does not involve any removed or added atoms. Four hexagons are transformed into two pentagons and two heptagons [SW(55-77) defect] by rotating one C-C bond by 90°. The existing SW defect was observed in recent experimental research [29], as shown in Figure 2a. The estimated formation energy (E_f) for SW(55-77) defect is 5eV by density functional theory (DFT) calculation [30,31], and 20 eV by molecular dynamics (MD) simulation [11]. Besides these atomic simulations, a topological continuum framework was proposed to evaluate the formation energy of associated and dissociated SW defects in graphene [32]. The high formation energy indicates a negligible kinetic formation rate of SW defect below 1000 °C. In addition, it has been reported that low mechanical strain (less than failure strain) cannot lead to the formation of SW defects [11].

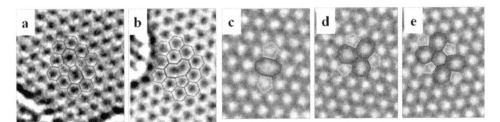

Figure 2. (a) Stone-Wales (SW55-77) defect, (b) Single vacancy (SV) defect ([29]. Reprinted with permission from American Chemical Society, Copyright 2008), (c) double vacancies (DV5-8-5), (d) double vacancies (DV555-777), (e) double vacancies (DV5555-6-7777) defect (Reprinted with permission from Ref [37] Copyright 2010 American Chemical Society).

Besides SW defect, another simple defect in graphene is missing lattice atoms. Single vacancy (SV) in graphene was experimentally observed using TEM [29,33] and scanning tunnelling microscope (STM) [34]. As shown in Figure 2b, one dangling bond remains toward the missing atom, which leads to the formation of a five-member ring and a nine-member ring. Such SV defect has formation energy $E_f \approx 7.5$ eV [35], which is much higher than that in many other materials (i.e. less than 3.0 eV in most metals). Double vacancies (DV) can also be created either by the combination of two SVs or by removing two neighbouring atoms. As shown in Figure 2c, two pentagons and one octagon (DV(5-8-5) defect) appear instead of four hexagons. Simulations [35] show that the formation energy of a DV (about 8 eV) is of the same order for a SV. In fact, the DV(5-8-5) is not even the energetically favour one. There are also other possible ways for a graphene lattice to arrange two missing atoms. For example, one C-C bond in the octagon of DV(5-8-5) defect transforms it into three pentagons and three heptagons (DV(555-777) defect) (Figure 2d). After rotating another C-C bond, DV (555-777) defect is transformed into DV(5555-6-7777) defect (Figure 2e). Multiple vacancies (MV) are created by removing more than 2 atoms. Generally, DV with even number of missing atoms are energetically favoured than that with odd number of missing atoms, where a dangling bond exists in the vicinity of defect [36].

Line defects

One-dimensional line defects have been observed in recent experimental studies [27,38,39]. Generally, these line defects have tilted boundaries separating two domains of different lattice orientations [37]. For example, a domain boundary has been observed to appear due to lattice mismatch in graphene grown on a Ni surface [27]. It is well-known that the properties of polycrystalline materials are governed by the size of grains as well as the atomic structure of grain boundaries, especially in two-dimensional graphene. In particular, grain boundaries may dominate the electronic transport in graphene [40].

2.3. Morphology

Generally, it is believed that long-range order does not exist according to Mermin-Wagner theorem [41]. Thus, dislocation should appear in 2D crystals at any finite temperature.

However, over the past two decades, researchers have demonstrated that long-range order can present due to anharmonic coupling between bending and stretching modes [42,43]. As a result, 2D membranes can exist but tend to be rippled. The typical height of roughness fluctuation scales with sample size L as L^ξ, with $\xi \approx 0.6$. Indeed, ripples in freestanding graphene were observed in recent experiments [44,45]. This kind of geometrical feature is generally referred as *intrinsic* morphology. In contrast, the morphology of substrate-supported graphene is regulated by the graphene-substrate interaction and is referred as *extrinsic* morphology. In this section, both *intrinsic* and *extrinsic* morphologies of graphene are reviewed.

Intrinsic morphology

As mentioned above, the shape of 2D graphene in 3D space is affected by its random intrinsic corrugations. The out-of-plane corrugations lead to increased strain energy but stabilize the random thermal fluctuation [46]. TEM observation indicates that suspended graphene sheets are generally not flat and the surface roughness can reach to about 1 nm [44]. In atomic force microscopy (AFM) measurements, nanometre-high buckles were observed in a single layer of graphene. The buckles in multi-layer graphene can penetrate from one layer to another [45]. To verify the experimental observation, simulations has been conducted to investigate the morphology of graphene and good agreement with the experiment has been achieved [47,48]. Atomistic Monte Carlo simulations also indicates that thermal fluctuation can create ripples with a ridge length around 8 nm [47], which is compatible with experimental findings (5-10 nm) [44].

Besides the effect of thermal fluctuation, sample size, aspect ratio, free edges and structural defects can also significantly affect the intrinsic morphologies of graphene. The constraint condition at the edge (e.g. periodic boundary or open edge) also affects the out-of-plane displacement [49]. As the aspect ratio (n) increases, its morphology changes from planar membrane, worm-like nanoribbons, and above the critical value $n_{cr}=50$, the nanoribbons self-fold into nanoscrolls, forming another structural phase, as shown in Figure 3. This implies that low aspect ratio in graphene nanoribbons is preferred for electronic applications as self-folding can be avoided.

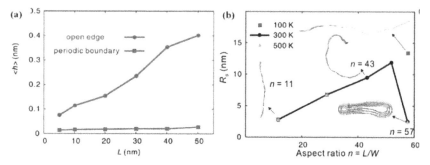

Figure 3. (a) Averaged out-of-plane displacement amplitude <h> of both graphene sheets with periodic boundary condition (red line) and open edges (blue line), and (b) Dependence of graphene sheet conformation on aspect ratio n=L/W (Reprinted with permission from Ref. [49] Copyright 2010 American Chemical Society).

For finite-sized graphene with open edges, the reconstruction of free edges results in non-zero edge stress. For regular (armchair and zigzag) and reconstructed edges terminated with hydrogen (r-H edge), they are subjected to compressive stresses [50-52]. Corresponding to the compressive stress, out-of-plane ripples are primarily confined to the edge areas. The influence of edge stresses is more dramatic in the nanoribbons than that in the sheets. Tensile stress is often associated with reconstructed edges terminated with pentagons-hexagons ring (r-5-6 edge) and pentagons-heptagons ring (r-5-7 edge)[53]. Such edge stress leads to large-scale curling of graphene sheets into cylindrical surfaces with their ends arching inward. Furthermore, attached chemical groups on graphene surface can change its morphology as a result of bond transition from sp^2 type to sp^3 type [54].

Extrinsic morphology

Graphene is also found to appear corrugations when fabricated on a substrate, which is often referred as the intrinsic morphology of graphene. Recent experiments indicate that unwanted photo-resist residue under the graphene can lead to such random corrugations. After removal of the resist residue, atomic-resolution images of graphene show that the graphene corrugations stem from its partial conformation to its substrate [55]. In addition, it has been demonstrated that single and few-layer graphene partially follow the surface morphology of the substrates [56-58]. These experimental studies suggest that the regulated extrinsic morphology of substrate-supported graphene is essentially different from that of free-standing graphene.

In terms of energy, the extrinsic morphology of graphene regulated by the supporting substrate is governed by the interplay among three types of energetics: (1) graphene strain energy, (2) graphene-substrate interaction energy and (3) substrate strain energy [46]. As graphene conforms to a substrate, the strain energy in the graphene and substrate increases but the graphene-substrate interaction energy decreases. By minimizing the total energy of the system, the equilibrium extrinsic morphology can be determined. In practice, the underlying substrate can be patterned with different features such as nanowires (1D), nanotubes (1D) or nanoparticles (0D). Graphene on a patterned substrate will conform to a regular extrinsic morphology.

For the substrate with 1D periodic sinusoidal surface, the regulated graphene is expected to have a similar morphology that can be described by

$$w_g(x) = A_g \cos\left(\frac{2\pi x}{\lambda}\right), \quad w_s(x) = A_s \cos\left(\frac{2\pi x}{\lambda}\right) - h \tag{1}$$

where λ is the wavelength; h is the distance between the middle planes of the graphene and the substrate surface; A_g and A_s are the amplitudes of the graphene morphology and the substrate surface, respectively. The graphene-substrate interaction energy is given by summing up all interaction energies between carbon and the substrate atoms via van der Waals force, i.e., [59]

$$E_{int} = \int_S \int_{V_S} V(r) \rho_S dV_S \rho_C dS \tag{2}$$

The strain energy of graphene sheet is given by

$$E_g = \frac{1}{\lambda/2}\int_0^{\lambda/2} D\Big/2\left(\frac{\partial^2 w_g}{\partial x^2}\right)^2 dx = \frac{4\pi^4 D A_g^2}{\lambda^4} \tag{3}$$

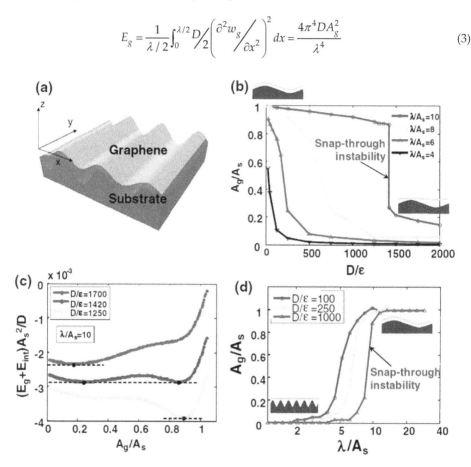

Figure 4. (a) Schematics of a graphene sheet on the corrugated substrate. (b) and (d) The normalized equilibrium amplitude of the graphene corrugation A_g/A_s as a function of D/ε for various λ/A_s. (c) Normalized total energy as a function of A_g/A_s for various D/ε (Reprinted with permission from Ref. [59] Copyright 2010 IOP Publishing).

In terms of the minimum potential energy, there exists a minimum value of $(E_g + E_s)$ where A_g and h define the equilibrium morphology of the graphene on the substrate. Figure 4 shows the normalized equilibrium amplitude of the graphene corrugation A_g/A_s as a function of D/ε for various λ/A_s. By analysing given substrate surface roughness (λ/A_s) and graphene-substrate interfacial bonding (D/ε) respectively, it was found that there is a sharp transition in the normalized equilibrium amplitude of the graphene corrugation. Such snap-through instability of the extrinsic morphology of graphene on the substrate can be understood by the energetic parameter shown in Figure 4c. Besides the interfacial bonding energy, the substrate surface roughness can also influence the extrinsic morphology graphene, as

shown in Figure 4d. Similar to the effect of substrate on the morphology of mounted graphene, 0D and 1D patterned nanoscale array can determine the equilibrium extrinsic morphology of graphene on the substrate [60,61].

3. Graphene-polymer nanocomposites

Polymer matrix nanocomposites with graphene and its derivatives as fillers have shown a
great potential for various important applications, such as electronics, green energy,
aerospace and automotiveindustries. As mentioned before, 2-D graphene possesses better
electrical, mechanical and thermal properties as well as other unique features, including
higher aspect ratio and larger specific surface area as compared to other reinforcements such
as CNTs and carbon and Kevlar fibres. It is reasonable to expect some significant
improvement in a range of properties in the composites with graphene as nanofiller. The
recent success in synthesis of large amount of graphene further promotes the development
of graphene based composite and hybrid materials.

3.1. Synthesis of graphene-polymer nanocomposites

Similar to processing other polymer matrix composites, solution blending, melt mixing and
in-situ polymerization are the commonly used approaches to produce graphene-polymer
composites.

Solution blending

Solution blending is the most popular technique to fabricate polymer-based composites in
that the polymer is readily soluble in common aqueous and organic solvents, such as water,
acetone, dimethylformamide (DMF), chloroform, dichloromethane (DCM) and toluene. This
technique includes the solubilisation of the polymer in suitable solvents, and mixing with
the solution of the dispersed suspension of graphene or graphene oxide (GO) platelets. The
polymers including PS [2], polycarbonate [62], polyacrylamide , polyimides [63] and
poly(methyl methacrylate) (PMMA) [64] have been successfully mixed with GO in solution
blending where the GO surface was usually functionalized using isocyanates, alkylamine
and alkyl-chlorosilanes to enhance its dispensability in organic solvents. In addition, the
facile production of aqueous GO platelet suspensions via sonication makes this technique
particularly appealing for water-soluble polymers such as poly(vinyl alcohol) (PVA) [65]
and poly(allylamine), composites of which can be produced via simple filtration [65].

For solution blending methods, the extent of exfoliation of GO platelets usually governs the
dispersion of GO platelets in the composite. Thus, solution blending offers a promising
approach to dispersing GO platelets into certain polymer matrix. Specifically, small molecule functionalization and grafting-to/from methods have been reported to achieve stable
GO platelet suspensions prior to mixing with polymer matrix. Some techniques, including
Lyophilizations methods [66], phase transfer techniques [67], and surfactants [68] have been
employed to facilitate solution blending of graphene-polymer nanocomposites. Nevertheless, surfactants may deteriorate composite properties. For example, the matrix-filler interfa-

cial thermal resistance in SWNT/polymer nanocomposites was increased by employing surfactants [69].

Melt mixing

Melt mixing technique utilizes a high temperature and shear forces to disperse the fillers in the polymer matrix. This process prevents the use of toxic solvents. Furthermore, compared with solution blending, melt mixing is often believed to be more cost effective. For graphene-polymer nanocomposites, the high temperature liquefies the polymer phase and allows easy dispersion or intercalation of GO platelets. However, the melt mixing is less effective in dispersing graphene sheets compared to solvent blending or in situ polymerization due to the increased viscosity at a high filler loading. The process can be applicable to both polar and non-polar polymers. Various graphene-based nanocomposites such as, exfoliated graphite–PMMA, graphene–polypropylene (PP), GO-poly (ethylene-2, 6-naphthalate) (PEN) and graphene–polycarbonate, can be fabricated by this technique. Even though the utility of graphene nanofiller is constrained by the low throughput of chemically reduced graphene in the melt mixing process, graphene production in bulk quantity in thermal reduction can be an appropriate choice for industrial scale production. However, the loss of the functional group in thermal reduction may be an obstacle in obtaining homogeneous dispersion in polymeric matrix melts especially in non-polar polymers.

In situ polymerization

This fabrication technique starts with mixing of filler in neat monomer (or multiple monomers), followed by polymerization in the presence of the dispersed filler. Then, precipitation/extraction or solution casting follows to generate samples for testing. In situ polymerization methods have produced composites with covalent crosslink between the matrix and filler. In addition, in situ polymerization has also produced non-covalent composites of a variety of polymers, such as poly (ethylene), PMMA and poly (pyrrole).

Unlike solution blending or melt mixing techniques, in situ polymerization technique achieves a high level of dispersion of graphene-based filler without prior exfoliation. It has been reported that monomer is intercalated between the layers of graphite or GO, followed by polymerization to separate the layers. This technique has been widely investigated for graphite or GO-derived polymer nanocomposites. For example, graphite can be intercalated by an alkali metal and a monomer, followed by polymerization initiated by thenegatively charged graphene sheets [70]. Although the polymerization may exfoliate the graphite nanoplatelets (GNPs), single-layer graphene platelets were not observed. TEM observation showed 3.6 nm thickness of graphene platelets with relatively low aspect ratio of about 30 dispersed in the PE matrix [71].

3.2. Fundamental properties

Mechanical properties

Higher mechanical properties of graphene sheets have attracted increasing attention worldwide. Similar to other composites, the mechanical properties depend on the concentration, aspect ratio and distribution of the nanofiller in the matrix and the interface bonding. For example, at a nanofiller weight fraction of 0.1±0.002%, the graphene-epoxy nanocompositesshow noticeable enhancement in the mechanical properties [3-5]. The Young's modulus, fracture strength of the nanocomposites are about ~31% and ~40% greater than the pristine epoxy, more efficient than the composites reinforced by multi-walled CNTs.The increase in fracture strength of the nanocomposites (graphene-PS) with 0.9 wt% graphene sheets is attributed to effective load transfer between the graphene layers and polymer matrix [72].

Besides simple reinforcing effects (Young's modulus and fracture strength), improvements in fracture toughness, fatigue strength and buckling resistance have also been reported in graphene-polymer nanocomposites[3,5,73-75]. For example, in situ polymerized graphene-epoxy nanocomposites show much higher buckling strength, fracture energy and fatigue strength than single- or multi-walled carbon nanotube-epoxy nanocomposites.However, the underlying strengthening and toughening mechanisms are still not well understood. Several factors, such as interfacial adhesion, spatial distribution and alignment of graphene nano-filler are considered to be crucial for effective reinforcement in the nanocomposites. Benefit-ing from improved interfacial adhesion, 76% and 62% the increase in elastic modulus and strength were achieved in the 0.7 wt% GO-PVA nanocomposite, respectively [76]. It was reported that the elastic modulus and fracture strength of Nylon-6 can be greatly improved by adding only 0.1 wt% GO [77]. The covalent bonding formed between the filler and matrix is attributed to the improved mechanical properties in the epoxy and polyurethane with GO-derived fillers [78-80]. Polymer nanocomposites with low loadings of functionalized graphene sheets (FGS) were reported to have a large shift in the glass transition temperature T_g[64]. In FGS-poly (acrylonitrile) nanocomposite, the shift in T_g is over 40°C when adding 1 wt% FGS filler loading. This behaviour can be attributed to the altered mobility of polymer chains at the filler-matrix interfaces [81,82]. Generally, a weak filler-matrix interface can constrain the chain mobility and thus increase the T_g.

Electrical properties

One of the most fascinating properties of graphene is its excellent electrical conductivity. When used as fillers in an insulating polymer matrix, conductive graphene may greatly improve the electrical conductivity of the composite. When the added graphene loading exceeds the electrical percolation threshold, a conductive network is expected in the poly-mer matrix. The conductivity σ_c as a function of filler loading can be described using a sim-ple power-law expression, i.e.,

$$\sigma_c = \sigma_f \left(\phi - \phi_c \right)^t \tag{4}$$

where ϕ is the filler volume fraction; ϕ_c is the percolation threshold; σ_f is the filler conductivity, and t is a scaling exponent. The overall electrical performance is dependent on the processing and dispersion, aggregation and alignment of the filler. The intrinsic characteristics of the filler such as aspect ratio and morphology also play a role in dictating the conductivity. The inter-sheet junction formed may affect the conductivity as well.

A high level of dispersion may not necessarily promote the onset of electrical percolation [83]. The polymer reason is a thin layer of polymer may coat on the well-dispersed fillers and prevent direct the formation of a conductive network. In fact, the lowest electrical percolation threshold for graphene-polymer nanocomposites was reported for the nanocomposite with heterogeneously dispersed graphene (about 0.15 wt %) [84]. For example, compression moulded polycarbonate and GO-polyester nanocomposites with aligned platelets showed an increased percolation threshold that isabout twice as high as the annealed samples with randomly oriented platelets [85,86]. Therefore, slight aggregation of the filler may lower the percolation threshold and improve the electrical conductivity of these nanocomposites [87]. Both theoretical analysis [88,89] and experiments demonstrated that the electrical conductivity of the nanocomposites correlates strongly with the aspect ratio of the platelets and higher aspect ratio leads to a higher conductivity. On the other hand, wrinkled, folded, or other non-flat morphologies may increase the electrical percolation threshold [90].

Thermal properties

The exceptional thermal properties of graphene have been used to improve the thermal stability and conductivity in nanocomposites. The 2D geometry of graphene-base materials may offer lower interfacial thermal resistance and thus provide higher thermal conductivity in the nano-composites. The 2D geometry of graphene also introduces anisotropy into the thermal conduc-tivity of graphene-polymer nanocomposites. For instance, the measured in-plane thermal con-ductivity is as much as ten times higher than the cross-plane conductivity [91]. Generally, ther-mal conductivity in nanocomposites can be analysed by percolation theory. Since phonons are the major mode of thermal conduction in amorphous polymers, covalent bonding between the filler and matrix can reduce phonon scattering at the filler-matrix interface, and subsequently enhance the thermal conductivity of nanocomposites [92]. In recent research, significant en-hancements in thermal conductivity have been achieved in graphene-epoxy nanocomposites, with conductivities increasing to 3~6 W/mK from 0.2 W/mK for neat epoxy. However, such significant improvement often needs a high filler loading, about 20 wt% and even higher. Some research has also reported improvement in thermal stability of graphene-polymer nanocompo-sites [93,94]. Furthermore, the negative coefficient of thermal expansion (CTE) [95] and high surface of graphene can lower the CTE of polymer matrix [96]. For instance, the CTE of a GO-epoxy nanocomposite with 5% filler loading decreases by nearly 32% for temperature below the polymer glass transition Temperature (T_g) [97].

4. Structure-property relationship

4.1. Microstructure effect

TEM and wide-angle X-ray scattering (WAXS), are often utilized to examine the dispersion of graphene fillers in composites. Sometimes, the morphological features of dispersed fillers can be missed out due to the tiny platelet thickness and intensity scattering. Recently, small-angle X-ray scattering (SAXS) and ultra-small-angle X-ray scattering (USAXS) have been increasingly used to examine the aggregation of filler at large material length scale.

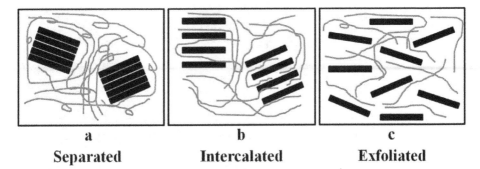

a	b	c
Separated	**Intercalated**	**Exfoliated**

Figure 5. Filler dispersion in graphene-based nanocomposites: (a) separated, (b) intercalated, and (c) exfoliated phases.

In GO-derived and GNP-polymer nanocomposites, the fillers can exist in different forms such as stacked, intercalated or exfoliated, as shown in Figure 5. As compared with the separated phase, increased interlayer spacing (in the order of few nanometres) can be achieved in the intercalated structures. In the exfoliated structure, exfoliated platelets have the largest interfacial contact with the polymer matrix, generally ideal for improvement of various properties of the composites. Due to increased interaction with the polymer matrix, exfoliated phase normally has a curved shape when embedded into a polymer matrix. The rumpled shape of filler then can result in mechanical interlocking, which is one the possible strengthening mechanisms. However, low modulus was also observed in the composite with wrinkled platelets [6]. The material processing methods can also influence the microstructure in nanocomposites. Randomly oriented exfoliated platelets can be achieved using solution blending or in situ polymerization [78]. Platelet restacking or incomplete exfoliation can also result in lower modulus due to decreased aspect ratio.

4.2. Interfacial behaviour

Although graphene-polymer nanocomposites exhibit excellent mechanical properties, the underlying strengthening and toughening mechanisms have not been well understood. Generally, it is believed that interfacial adhesion plays a key role in determining the improvements in mechanical properties of graphene-polymer nanocomposites. Low interfacial adhesion may lead to lower load transfer between the filler and matrix. Both AFM and Raman spectroscopy [98,99] can be utilized to measure the graphene-polymer interfacial adhesion. Raman spectra and their Raman bands were found to shift with stress, which enables stress-transfer to be monitored between the matrix and reinforcing phase. Moreover, a universal calibration has been established between the rate of shift of the G′ band with strain[100]. Recently, interfacial shear stress [98] and effective Young's modulus [101] were successfully determined using Raman spectroscopy. The relationship between matrix strain$\varepsilon_m$and strain in the graphene flake$\varepsilon_f$can be described by

$$\varepsilon_f = \varepsilon_m \left[1 - \frac{\cosh\left(ns\frac{x}{l}\right)}{\cosh\left(ns/2\right)} \right] \tag{5}$$

$$n = \sqrt{\frac{2G_m}{E_f}} \tag{6}$$

where n and G_m is the matrix shear modulus, E_f is the Young's modulus of the graphene filler. The variation of shear stress τ_i, at the graphene-polymer interface is given by

$$\tau_i = nE_f\varepsilon_m \frac{\sinh\left(ns\frac{l}{x}\right)}{\cosh\left(ns/2\right)} \tag{7}$$

Corresponding to ε_m=0.4%, there is a good agreement between the measured and predicted (Equation 5) variation of fibre strain with position on the monolayer, validating the use of the shear lag analysis. At ε_m=0.6%, however, the interface failure occurs between the filler and polymer and stress transfer is taking place through interfacial friction. The interfacial shear stress (ISS) between graphene and polymer is determined in the range of ~0.3-0.8 MPa, much lower than that between CNTs and polymer (~20-40 MPa). The low ISS was attributed to the poor interface adhesion.

Raman spectroscopy analysis [98,102] confirmed the reinforcement effect of graphene and its dependence with crystallographic orientations and the layer number of graphene. It is demonstrated that monolayer or bilayer graphene has better load transfer than tri-layer or multi-layer [103]. Without compromising an even dispersion, higher filler loading is easy to achieve with multilayer graphene. There is therefore a balance in design of graphene-polymer nanocomposites between a higher filler loading and decreased load transfer as the number of layers in graphene filler increases. Raman G-band analysis suggested that load transfer at the GPL-PDMS interface is more effective in comparison to that along single wall carbon nanotube/PDMS interface [7]. In terms of loading mode, it is interesting to note that the GPL fillers went into compression under tensile loading and vice versa. Up to now, interface load transfer, mechanical interlocking caused by wrinkled surface and defects in graphene are main factor in controlling the reinforcement mechanisms. Due to the complex interactions between graphene, functional groups attached and the polymer, controversial results are often observed in the load transfer analysis. Further theoretical and numerical analysis is much needed.

5. Conclusion

In summary, the interesting properties of graphene and its composites mentioned above have led to the exploration of numerous applications such as transistors, chemical and biosensors, energy storage devices, nanoelectro-mechanical systems and others, just as the research community has done with carbon nanotubes previously. The past decade has witnessed the rapid growth of carbon-based nanotechnology. Further research in the area will assist the development of next generation graphene based composites and hybrid materials.

Author details

Mingchao Wang, Cheng Yan and Lin Ma
School of Chemistry, Physics and Mechanical Engineering, Science and Engineering Faculty, Queensland University of Technology, Brisbane, Australia

6. References

[1] Novoselov KS, Geim AK, Morozov SV, Jiang D, Zhang Y, Dubonos SV, Grigorieva IV, Firsov AA, Science. 306 (2004).
http://www.sciencemag.org/content/306/5696/666.abstract.

[2] Stankovich S, Dikin DA, Dommett GHB, Kohlhaas KM, Zimney EJ, Stach EA, Piner RD, Nguyen ST, Ruoff RS, Nature. 442 (2006).
http://dx.doi.org/10.1038/nature04969.

[3] Rafiee MA, Rafiee J, Wang Z, Song H, Yu Z-Z, Koratkar N, ACS Nano. 3 (2009).
http://dx.doi.org/10.1021/nn9010472.

[4] Rafiee MA, Lu W, Thomas AV, Zandiatashbar A, Rafiee J, Tour JM, Koratkar NA, ACS Nano. 4 (2010).
http://dx.doi.org/10.1021/nn102529n.

[5] Rafiee MA, Rafiee J, Srivastava I, Wang Z, Song HH, Yu ZZ, Koratkar N, Small. 6 (2010).
http://dx.doi.org/10.1002/smll.200901480.

[6] Wakabayashi K, Pierre C, Dikin DA, Ruoff RS, Ramanathan T, Brinson LC, Torkelson JM, Macromolecules. 41 (2008).
http://dx.doi.org/10.1021/ma071687b.

[7] Srivastava I, Mehta RJ, Yu Z-Z, Schadler L, Koratkar N, Appl. Phys. Lett. 98 (2011).
http://link.aip.org/link/?APL/98/063102/1.

[8] Lee C, Wei X, Kysar JW, Hone J, Science. 321 (2008).
http://www.sciencemag.org/content/321/5887/385.abstract.

[9] Scharfenberg S, Rocklin DZ, Chialvo C, Weaver RL, Goldbart PM, Mason N, Appl. Phys. Lett. 98 (2011).
http://link.aip.org/link/?APL/98/091908/1.

[10] Zhao H, Min K, Aluru NR, Nano Lett. 9 (2009).
http://dx.doi.org/10.1021/nl901448z.

[11] Wang MC, Yan C, Ma L, Hu N, Chen MW, Comp. Mater. Sci. 54 (2012).
http://www.sciencedirect.com/science/article/pii/S0927025611006033.

[12] Zhao H, Aluru NR, J. Appl. Phys. 108 (2010).
http://link.aip.org/link/?JAP/108/064321/1.

[13] Tsoukleri G, Parthenios J, Papagelis K, Jalil R, Ferrari AC, Geim AK, Novoselov KS, Galiotis C, Small. 5 (2009).

http://dx.doi.org/10.1002/smll.200900802.

[14] Lee C, Wei X, Li Q, Carpick R, Kysar JW, Hone J, physica status solidi (b). 246 (2009). http://dx.doi.org/10.1002/pssb.200982329.

[15] Koenig SP, Boddeti NG, Dunn ML, Bunch JS, Nat Nano. 6 (2011). http://dx.doi.org/10.1038/nnano.2011.123.

[16] Zhang Y, Tan Y-W, Stormer HL, Kim P, Nature. 438 (2005). http://dx.doi.org/10.1038/nature04235.

[17] Novoselov KS, Jiang Z, Zhang Y, Morozov SV, Stormer HL, Zeitler U, Maan JC, Boebinger GS, Kim P, Geim AK, Science. 315 (2007). http://www.sciencemag.org/content/315/5817/1379.abstract.

[18] Geim AK, Novoselov KS, Nat. Mater. 6 (2007). http://dx.doi.org/10.1038/nmat1849.

[19] Somani PR, Somani SP, Umeno M, Chem. Phys. Lett. 430 (2006). http://www.sciencedirect.com/science/article/pii/S0009261406009018.

[20] Bae S, Kim H, Lee Y, Xu X, Park J-S, Zheng Y, Balakrishnan J, Lei T, Ri Kim H, Song YI, Kim Y-J, Kim KS, Ozyilmaz B, Ahn J-H, Hong BH, Iijima S, Nat Nano. 5 (2010). http://dx.doi.org/10.1038/nnano.2010.132.

[21] Kim KS, Zhao Y, Jang H, Lee SY, Kim JM, Kim KS, Ahn J-H, Kim P, Choi J-Y, Hong BH, Nature. 457 (2009). http://dx.doi.org/10.1038/nature07719.

[22] Berger C, Song Z, Li T, Li X, Ogbazghi AY, Feng R, Dai Z, Marchenkov AN, Conrad EH, First PN, de Heer WA, The Journal of Physical Chemistry B. 108 (2004). http://dx.doi.org/10.1021/jp040650f.

[23] Berger C, Song Z, Li X, Wu X, Brown N, Naud C, Mayou D, Li T, Hass J, Marchenkov AN, Conrad EH, First PN, de Heer WA, Science. 312 (2006). http://www.sciencemag.org/content/312/5777/1191.abstract.

[24] Zheng Q, Geng Y, Wang S, Li Z, Kim J-K, Carbon. 48 (2010). http://www.sciencedirect.com/science/article/B6TWD-50NBNTH-1/2/f9d91c8358333a00160e505b972e0cd3.

[25] Pei QX, Zhang YW, Shenoy VB, Carbon. 48 (2010). http://www.sciencedirect.com/science/article/B6TWD-4XNF8DY-3/2/d59770660b4ae737f9385b89c3439a0c.

[26] OuYang F, Huang B, Li Z, Xiao J, Wang H, Xu H, J. Phys. Chem. C. 112 (2008). http://dx.doi.org/10.1021/jp710547x.

[27] Lahiri J, Lin Y, Bozkurt P, Oleynik II, Batzill M, Nat Nano. 5 (2010). http://dx.doi.org/10.1038/nnano.2010.53.

[28] Stone AJ, Wales DJ, Chem. Phys. Lett. 128 (1986). http://www.sciencedirect.com/science/article/pii/0009261486806613.

[29] Meyer JC, Kisielowski C, Erni R, Rossell MD, Crommie MF, Zettl A, Nano Lett. 8 (2008). http://dx.doi.org/10.1021/nl801386m.

[30] Li L, Reich S, Robertson J, Phys. Rev. B. 72 (2005). http://link.aps.org/doi/10.1103/PhysRevB.72.184109.

[31] Ma J, Alfè D, Michaelides A, Wang E, Phys. Rev. B. 80 (2009).

http://link.aps.org/doi/10.1103/PhysRevB.80.033407.

[32] Ertekin E, Chrzan DC, Daw MS, Phys. Rev. B. 79 (2009).
http://link.aps.org/doi/10.1103/PhysRevB.79.155421.

[33] Gass MH, Bangert U, Bleloch AL, Wang P, Nair RR, Geim AK, Nat Nano. 3 (2008).
http://dx.doi.org/10.1038/nnano.2008.280.

[34] Ugeda MM, Brihuega I, Guinea F, Gómez-Rodríguez JM, Phys. Rev. Lett. 104 (2010).
http://link.aps.org/doi/10.1103/PhysRevLett.104.096804.

[35] Krasheninnikov AV, Lehtinen PO, Foster AS, Nieminen RM, Chem. Phys. Lett. 418
(2006).
http://www.sciencedirect.com/science/article/pii/S0009261405016544.

[36] Kotakoski J, Krasheninnikov AV, Nordlund K, Phys. Rev. B. 74 (2006).
http://link.aps.org/doi/10.1103/PhysRevB.74.245420.

[37] Banhart F, Kotakoski J, Krasheninnikov AV, ACS Nano. 5 (2010).
http://dx.doi.org/10.1021/nn102598m.

[38] Coraux J, N`Diaye AT, Busse C, Michely T, Nano Lett. 8 (2008).
http://dx.doi.org/10.1021/nl0728874.

[39] Cervenka J, Katsnelson MI, Flipse CFJ, Nat Phys. 5 (2009).
http://dx.doi.org/10.1038/nphys1399.

[40] Yazyev OV, Louie SG, Phys. Rev. B. 81 (2010).
http://link.aps.org/doi/10.1103/PhysRevB.81.195420.

[41] Mermin ND, Physical Review. 176 (1968).
http://link.aps.org/doi/10.1103/PhysRev.176.250.

[42] Nelson DR, Peliti L, Journal De Physique. 48 (1987).
http://hal.archives-ouvertes.fr/jpa-00210530/en/.

[43] Le Doussal P, Radzihovsky L, Phys. Rev. Lett. 69 (1992).
http://link.aps.org/doi/10.1103/PhysRevLett.69.1209.

[44] Meyer JC, Geim AK, Katsnelson MI, Novoselov KS, Booth TJ, Roth S, Nature. 446 (2007).
http://dx.doi.org/10.1038/nature05545.

[45] Li Z, Cheng Z, Wang R, Li Q, Fang Y, Nano Lett. 9 (2009).
http://dx.doi.org/10.1021/nl901815u.

[46] Teng L, Model. Simul. Mater. Sc. 19 (2011).
http://stacks.iop.org/0965-0393/19/i=5/a=054005.

[47] Fasolino A, Los JH, Katsnelson MI, Nature Materials. 6 (2007).
http://www.nature.com/nmat/journal/v6/n11/full/nmat2011.html.

[48] Duan WH, Gong K, Wang Q, Carbon. 49 (2011).
http://www.sciencedirect.com/science/article/pii/S0008622311002247.

[49] Xu Z, Buehler MJ, ACS Nano. 4 (2010).
http://dx.doi.org/10.1021/nn100575k.

[50] Shenoy VB, Reddy CD, Ramasubramaniam A, Zhang YW, Phys. Rev. Lett. 101 (2008).
http://link.aps.org/doi/10.1103/PhysRevLett.101.245501.

[51] Huang B, Liu M, Su N, Wu J, Duan W, Gu B-l, Liu F, Phys. Rev. Lett. 102 (2009).
http://link.aps.org/doi/10.1103/PhysRevLett.102.166404.

[52] Bets KV, Yakobson BI, Nano Research. 2 (2009).

http://www.nanoarchive.org/8385/.

[53] Reddy CD, Ramasubramaniam A, Shenoy VB, Zhang Y-W, Appl. Phys. Lett. 94 (2009).
http://link.aip.org/link/?APL/94/101904/1
http://dx.doi.org/10.1063/1.3094878.

[54] Pei QX, Zhang YW, Shenoy VB, Nanotechnology. 21 (2010).
http://dx.doi.org/10.1088/0957-4484/21/11/115709.

[55] Ishigami M, Chen JH, Cullen WG, Fuhrer MS, Williams ED, Nano Lett. 7 (2007).
http://dx.doi.org/10.1021/nl070613a.

[56] Stolyarova E, Rim KT, Ryu S, Maultzsch J, Kim P, Brus LE, Heinz TF, Hybertsen MS,
Flynn GW, P. Natl. Acad. Sci. USA. 104 (2007).
http://www.pnas.org/content/104/22/9209.abstract.

[57] Stoberl U, Wurstbauer U, Wegscheider W, Weiss D, Eroms J, Appl. Phys. Lett. 93 (2008).
http://dx.doi.org/10.1063/1.2968310.

[58] Geringer V, Liebmann M, Echtermeyer T, Runte S, Schmidt M, Rückamp R, Lemme MC,
Morgenstern M, Phys. Rev. Lett. 102 (2009).
http://link.aps.org/doi/10.1103/PhysRevLett.102.076102.

[59] Teng L, Zhao Z, J. Phys. D Appl. Phys. 43 (2010).
http://stacks.iop.org/0022-3727/43/i=7/a=075303.

[60] Zhang Z, Li T, J. Appl. Phys. 107 (2010).
http://dx.doi.org/10.1063/1.3427551.

[61] Zhang Z, Li T, J. Nanomater. 2011 (2011).
http://dx.doi.org/10.1155/2011/374018.

[62] Higginbotham AL, Lomeda JR, Morgan AB, Tour JM, ACS Appl. Mater. Inter. 1 (2009).
http://dx.doi.org/10.1021/am900419m.

[63] Chen D, Zhu H, Liu T, ACS Appl. Mater. Inter. 2 (2010).
http://dx.doi.org/10.1021/am1008437.

[64] RamanathanT, Abdala AA, StankovichS, Dikin DA, Herrera Alonso M, Piner RD,
Adamson DH, Schniepp HC, ChenX, Ruoff RS, Nguyen ST, Aksay IA, Prud'Homme
RK, Brinson LC, Nat Nano. 3 (2008).
http://dx.doi.org/10.1038/nnano.2008.96.

[65] Xu Y, Hong W, Bai H, Li C, Shi G, Carbon. 47 (2009).
http://www.sciencedirect.com/science/article/pii/S0008622309005296.

[66] Cao Y, Feng J, Wu P, Carbon. 48 (2010).
http://www.sciencedirect.com/science/article/pii/S0008622310004549.

[67] Wei T, Luo G, Fan Z, Zheng C, Yan J, Yao C, Li W, Zhang C, Carbon. 47 (2009).
http://www.sciencedirect.com/science/article/pii/S0008622309002565.

[68] Lee HB, Raghu AV, Yoon KS, Jeong HM, J. Macromol. Sci. B. 49 (2010).
http://www.tandfonline.com/doi/abs/10.1080/00222341003603701.

[69] Bryning MB, Milkie DE, Islam MF, Kikkawa JM, Yodh AG, Appl. Phys. Lett. 87 (2005).
http://dx.doi.org/10.1063/1.2103398.

[70] Shioyama H, Synthetic Metals. 114 (2000).
http://www.sciencedirect.com/science/article/pii/S0379677900002228.

[71] Fim FdC, Guterres JM, Basso NRS, Galland GB, J. Polym. Sci. Pol. Chem. 48 (2010).

http://dx.doi.org/10.1002/pola.23822.

[72] Fang M, Wang K, Lu H, Yang Y, Nutt S, J. Mater. Chem. 19 (2009).
http://dx.doi.org/10.1039/B908220D.

[73] Rafiee MA, Rafiee J, Yu ZZ, Koratkar N, Appl. Phys. Lett. 95 (2009).
http://apl.aip.org/applab/v95/i22/p223103_s1?isAuthorized=no.

[74] Yavari F, Rafiee MA, Rafiee J, Yu ZZ, Koratkar N, ACS Appl. Mater. Inter. 2 (2010).
http://dx.doi.org/10.1021/am100728r.

[75] Rafiq R, Cai D, Jin J, Song M, Carbon. 48 (2010).
http://www.sciencedirect.com/science/article/pii/S0008622310005476.

[76] Liang J, Huang Y, Zhang L, Wang Y, Ma Y, Guo T, Chen Y, Adv. Funct. Mater. 19 (2009).
http://dx.doi.org/10.1002/adfm.200801776.

[77] Xu Z, Gao C, Macromolecules. 43 (2010).
http://dx.doi.org/10.1021/ma1009337.

[78] Kim H, Miura Y, Macosko CW, Chem. Mater. 22 (2010).
http://dx.doi.org/10.1021/cm100477v.

[79] Lee YR, Raghu AV, Jeong HM, Kim BK, Macromol. Chem. Physic. 210 (2009).
http://dx.doi.org/10.1002/macp.200900157.

[80] Miller SG, Bauer JL, Maryanski MJ, Heimann PJ, Barlow JP, Gosau J-M, Allred RE, Compos. Sci. Technol. 70 (2010).
http://www.sciencedirect.com/science/article/pii/S0266353810000825.

[81] Bansal A, Yang H, Li C, Cho K, Benicewicz BC, Kumar SK, Schadler LS, Nat. Mater. 4 (2005).
http://dx.doi.org/10.1038/nmat1447.

[82] Priestley RD, Ellison CJ, Broadbelt LJ, Torkelson JM, Science. 309 (2005).
http://www.sciencemag.org/content/309/5733/456.abstract.

[83] Schaefer DW, Justice RS, Macromolecules. 40 (2007).
http://dx.doi.org/10.1021/ma070356w.

[84] Pang H, Chen T, Zhang G, Zeng B, Li Z-M, Mater. Lett. 64 (2010).
http://www.sciencedirect.com/science/article/pii/S0167577X10005379.

[85] Kim H, Macosko CW, Macromolecules. 41 (2008).
http://dx.doi.org/10.1021/ma702385h.

[86] Kim H, Macosko CW, Polymer. 50 (2009).
http://www.sciencedirect.com/science/article/pii/S0032386109004558.

[87] Bauhofer W, Kovacs JZ, Compos. Sci. Technol. 69 (2009).
http://www.sciencedirect.com/science/article/pii/S026635380800239X.

[88] Hicks J, Behnam A, Ural A, Appl. Phys. Lett. 95 (2009).
http://dx.doi.org/10.1063/1.3267079.

[89] Li J, Kim J-K, Compos. Sci. Technol. 67 (2007).
http://www.sciencedirect.com/science/article/pii/S0266353806004386.

[90] Yi YB, Tawerghi E, Phys. Rev. E. 79 (2009).
http://link.aps.org/doi/10.1103/PhysRevE.79.041134.

[91] Veca LM, Meziani MJ, Wang W, Wang X, Lu F, Zhang P, Lin Y, Fee R, Connell JW, Sun Y-P, Adv. Mater. 21 (2009).
http://dx.doi.org/10.1002/adma.200802317.

[92] Ganguli S, Roy AK, Anderson DP, Carbon. 46 (2008).
http://www.sciencedirect.com/science/article/pii/S000862230800078X.

[93] Kim I-H, Jeong YG, Journal of Polymer Science Part B: Polymer Physics. 48 (2010).
http://dx.doi.org/10.1002/polb.21956.

[94] Bao Q, Zhang H, Yang J-x, Wang S, Tang DY, Jose R, Ramakrishna S, Lim CT, Loh KP, Adv. Funct. Mater. 20 (2010).
http://dx.doi.org/10.1002/adfm.200901658.

[95] Bao W, Miao F, Chen Z, Zhang H, Jang W, Dames C, Lau CN, Nat Nano. 4 (2009).
http://dx.doi.org/10.1038/nnano.2009.191.

[96] Paul DR, Robeson LM, Polymer. 49 (2008).
http://www.sciencedirect.com/science/article/pii/S0032386108003157.

[97] Wang S, Tambraparni M, Qiu J, Tipton J, Dean D, Macromolecules. 42 (2009).
http://dx.doi.org/10.1021/ma900631c.

[98] Gong L, Kinloch IA, Young RJ, Riaz I, Jalil R, Novoselov KS, Adv. Mater. 22 (2010).
http://onlinelibrary.wiley.com/doi/10.1002/adma.200904264/abstract.

[99] Kranbuehl DE, Cai M, Glover AJ, Schniepp HC, J. Appl. Pol. Sci. 122 (2011).
http://dx.doi.org/10.1002/app.34787.

[100] Cooper CA, Young RJ, Halsall M, Compos A Appl. Sci. Manu. 32 (2001).
http://www.sciencedirect.com/science/article/pii/S1359835X0000107X.

[101] Frank O, Tsoukleri G, Riaz I, Papagelis K, Parthenios J, Ferrari AC, Geim AK, Novoselov KS, Galiotis C, Nat Commun. 2 (2011).
http://dx.doi.org/10.1038/ncomms1247.

[102] Young RJ, Gong L, Kinloch IA, Riaz I, Jalil R, Novoselov KS, ACS Nano. 5 (2011).
http://pubs.acs.org/doi/abs/10.1021/nn2002079.

[103] Gong L, Young RJ, Kinloch IA, Riaz I, Jalil R, Novoselov KS, ACS Nano. (2012).
http://dx.doi.org/10.1021/nn203917d.

Chapter 2

Graphene-Boron Nitride Composite: A Material with Advanced Functionalities

Sumanta Bhandary and Biplab Sanyal

Additional information is available at the end of the chapter

1. Introduction

The discovery of two dimensional materials is extremely exciting due to their unique properties, resulting from the lowering of dimensionality. Physics in 2D is quite rich (e.g., high temperature superconductivity, fractional quantum Hall effect etc.) and is different from its other dimensional counterparts. A 2D material acts as the bridge between bulk 3D systems and 0D quantum dots or 1D chain materials. This can well be the building block for materials with other dimensions. The discovery of graphene, the 2D allotrope of carbon by Geim and Novoselov [1] made an enormous sensation owing to a plethora of exciting properties. They were awarded the Nobel prize in Physics in 2010. Graphene, an atomically thick C layer, has broken the jinx of impossibility of the formation of a 2D structure at a finite temperature, as argued by Landau *et al.*[2, 3]. The argument that a 2D material is thermodynamically unstable due to the out-of-plane thermal distortion, which is comparable to its bond length, was proven invalid with this discovery. One of the recent interests is to understand this apparent discrepancy by considering rippled structures of graphene at finite temperatures.

Graphene, with its exciting appearance, has won the crowns of the thinnest, the strongest, the most stretchable material along with extremely high electron mobility and thermal conductivity [4]. The linear dispersion curve at the Dirac point gives rise to exciting elementary electronic properties. Electrons in graphene behave like massless Dirac fermions, similar to the relativistic particles in quantum electrodynamics and hence has brought different branches of science together under a truly interdisciplinary platform. At low temperature and high magnetic field, a fascinating phenomenon, called the half integer quantum hall effect, is observed. The relativistic nature of carriers in graphene shows 100% tunneling through a potential barrier by changing its chirality. The phenomenon is known as "Klein-Paradox". Minimum conductivity of a value of conductivity quantum (e^2/h per spin per valley) is measured at zero field, which makes graphene unique. "The CERN on table top" is thus a significant naming of the experiments performed with this fascinating material [5].

An infinite pristine graphene is a semi metal, i. e., a metal with zero band gap [6]. Inversion symmetry provided by $P6/mmm$ space group results in a band degeneracy at the Dirac points

(K and K') in the hexagonal Brillouin zone (BZ). This limits its most anticipated application in electronics as the on-off current ratio becomes too small to be employed in a device. The opening of a band gap is thus essential from electronics point of view retaining a high carrier mobility. Several approaches have been already made by modifying graphene, either chemically [7, 8] or by structural confinement [9–11] to improve its application possibilities, both from theory and experiment. It should be noted that in theoretical studies, the use of density functional theory (DFT) [12] has always played an instrumental role in understanding and predicting the properties of materials, often in a quantitative way.

Boron Nitride (BN), on the other hand, can have different forms of structures like bulk hexagonal BN with sp^2 bond, cubic BN with sp^3 bond, analogous to graphite and diamond respectively. A 2D sheet with strong sp^2 bonds can also be derived from it, which resembles its carbon counterpart, graphene. But two different chemical species in the two sublattices of BN forbid the inversion symmetry, which results in the degeneracy lifting at Dirac points in the BZ. Hexagonal BN sheet thus turns out to be an insulator with a band gap of 5.97 eV.

This opens up a possibility of alloying these neighboring elements in the periodic table to form another interesting class of materials. Possibilities are bright and so are the promises. B-N bond length is just 1.7% larger than the C-C bond, which makes them perfect for alloying with minimal internal stress. At the same time, introduction of BN in graphene, breaks the inversion symmetry, which can result in the opening up of a band gap in graphene. On top of that, the electronegativities of B, C, and N are respectively 2.0, 2.5 and 3.0 [13], which means that the charge transfer in different kinds of BCN structures is going to play an interesting role both in stability and electronic properties.

Hexagonal BNC (h-BNC) films have been recently synthesized [14] on a Cu substrate by thermal catalytic chemical vapor deposition method. For the synthesis, ammonia borane (NH_3-BH_3) and methane were used as precursors for BN and C respectively. In the experimental situation, it is possible to control the relative percentage of C and BN. The interesting point is that the h-BNC films can be lithographically patterned for fabrication of devices. The atomic force microscopy images indicated the formation of 2-3 layers of h-BNC. The structures and compositions of the films were characterized by atomic high resolution transmission electron microscopy and electron energy-loss spectroscopy. Electrical measurements in a four-probe device showed that the electrical conductivity of h-BNC ribbons increased with an increase in the percentage of graphene. The h-BNC field effect transistor showed ambipolar behavior similar to graphene but with reduced carrier mobility of 5-20 $cm^2V^{-1}s^{-1}$. From all these detailed analysis, one could conclude that in h-BNC films, hybridized h-BN and graphene domains were formed with unique electronic properties. Therefore, one can imagine the h-BN domains as extended impurities in the graphene lattice.

The structure and composition of BN-graphene composite are important issues to consider. As mentioned before, substitution of C in graphene by B and N can give an alloyed BCN configuration. Considering the possibilities of thermodynamic non-equilibrium at the time of growth process, one can think of several ways of alloying. The potential barrier among those individual structures can be quite high and that can keep these relatively high energetic structures stable at room temperature. For example, a huge potential barrier has to be crossed to reach a phase segregated alloy from a normal alloy, which makes normal alloy stable at room temperature. Now, depending on the growth process, different types of alloying are possible. Firstly, one can think of an even mixture of boron nitride and carbon, where one C_2

block is replaced by B-N. In this case, the formula unit will be BC_2N. Secondly, a whole area of graphene can be replaced by boron nitride, which makes them phase separated. This we call as phase segregated alloy. The formula unit of phase segregated alloy can change depending on the percentage of doping. The final part of the following section will be devoted to the phase segregated BCN alloys. Apart from those, a distributed alloying is possible with different BN:graphene ratios.

The substitution of C_2 with B-N introduces several interesting features. Firstly, B-N bond length is 1.7% bigger that C-C bond but C-B bond is 15% bigger than C-N bond. So, this is going to create intra-layer strain, which is going to affect its stability. Secondly, the difference in electro negativity in B (2.0) and N (3.0) will definitely cause a charge transfer. The orientation of charged pair B-N do have a major contribution in cohesive energy. Thirdly, as mentioned earlier, this will break inversion symmetry in graphene, which brings a significant change in electronic properties. Keeping these in mind, we are now going to discuss stability and electronic structure of BC_2N.

2. Stability of BC_2N

In this section we will mainly focus on the stability issues for various BC_2N structures [13, 15]. To demonstrate the factors for structural stability, we have chosen five different structures of BC_2N (Fig. 1). Let's first have a closer look at structure I. Every C atom has one C, N and B as nearest neighbors while B(N) has two Cs and one N(B) as their nearest neighbors. There is a possibility of all bonds to be relaxed, retaining the hexagonal structure. Stress is thus minimized in this structure, which helps obviously in the stability. In structure II, each C has two C and either one B or N as nearest neighbors. C_2 and BN form own striped regions, which lie parallel to each other in this structure. Now C-B bond is much larger that C-C. So this is definitely going to put some internal stress. From the point of view of intra-layer stress, this structure is definitely less stable than the previous one. Looking at structure III, one can see this structure looks similar to structure I but B-N bond orientations are different. Each C atom now has either two N or B and one C in its neighboring position while N(B) has two C and one B (N) as neighbors. This obviously adds some uncompensated strain in the structure. Structure III, thus consists of two parallel C-N and C-B chains and as C-B bond length is much larger than C-N (15%), this mismatch is going to introduce a large strain in the interface. On the other hand in structure I, C-N and C-B are lined up making the structural energy lower compared to structure III. Structure IV does not contain any C-C bond. C-B and C-N chains are lined parallel to each other. Finally in structure V, B-N bonds are placed in such a way that they make $60°$ angle to each other. Both of the last two structures thus have uncompensated strain, which increases their structural energies.

Bond energy is another key factor in stability. When the bond energies are counted, the ordering of the bonds is the following [13]:

$$B - N(4.00\,eV) > C - C(3.71\,eV) > N - C(2.83\,eV) >$$
$$B - C(2.59\,eV) > B - B(2.32\,eV) > N - N(2.11\,eV)$$

The maximization of stable bonds like B-N and C-C will thus stabilize the structure as a whole. Now, a structure like II, with a striped pattern of C and B-N chains has maximum number of such bonds. This makes it most stable even though a structural strain is present. In this case bond energy wins over structural stress. For the structures like I and III, number

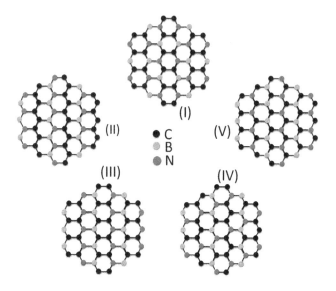

Figure 1. Crystal structure of different isomers of BC_2N. Filled black, pink, and blue circles represent carbon, boron and nitrogen respectively.

of such bonds is equal. In that case, intra-layer stress acts as the deciding factor. Structure IV, on the other hand does not have any C-C or B-N bond but only C-B or C-N. Therefore, the issue of stability is the most prominent here. The number of strong bonds is sufficiently large in structure V for the stabilization despite of $60°$ arrangement of B-N bonds.

Another important issue is the charge transfer as there is a difference between the electronegativities of B, C, and N. As mentioned earlier, N is the most electronegative and B is the least one while C behaves as a neutral atom. This also adds an ionic character in the bond formation. B (N) always gains some +ve (-ve) charges. So, the gain in the electrostatic energy only happens if these +/- charges are situated in an alternative manner. Otherwise the electrostatic repulsion makes the structure unstable. From this point of view, structures II, III, and V are more stable than the other two following the trend shown from bond energies. Thus, as reported by Itoh et al.[13], ordering of the structures will be: II>V>(I,III)>IV . Stability of other possible isomers can as well be anticipated with the same arguments.

So far we have talked about the substitutional alloying of BN and graphene, where BN to C_2 ratio is 1:1. Another group of structures, which can be formed by alloying BN and graphene is the phase segregated BCN. In this kind of structure, BN (graphene) retains its own phase, separated by graphene (BN). The experimental evidence of these kinds of structures have been shown [14]. The size of the graphene or BN phase has an impact on the stability and electronic properties. Here, the BN:C_2 ratio is thus not only 1:1 but can be varied and if varied controllably, one can control the electronic properties such as band gap [16]. Lam et al.[16] have shown that, by controlling the graphene phase, one can control the band gap according to the desired values for technological applications. The phase segregated $(BN)_m(C_2)_n$ alloys are also found to be stable over the first kind of alloying, which indicates a transformation due to thermal vibration. Yuge et al.[17] with DFT studies and Monte Carlo simulations have shown a tendency of phase separation between BN and graphene.

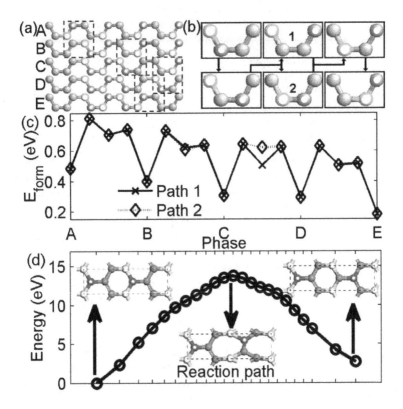

Figure 2. (a) Different steps (A-E) of phase separation process, (b) Swapping of BN and C dimers, (c) Formation energies for different steps shown in (a) for two different paths demonstrated in (b), (d) Activation energy in going from left to right configuration in the initial step of phase segregation. Reprinted with permission from Appl. Phys. Lett. 98, 022101 (2011). Copyright (2011) American Institute of Physics.

Even though a tendency is indicated, a recent calculation by Lam et al., have shown that this possibility is hindered as the activation energy required for phase segregation is extremely high. As shown in Fig. 2, they have chosen a possible path for phase separation by swapping B-N bond to C-C bonds. This kind of swapping can also happen in two ways (Fig. 2(b)). Calculated formation energies for these two process are shown in Fig. 2(c), which basically demonstrates that the intermediate structures are quite high in energy compared to evenly distributed and phase separated structures. The authors also performed nudged elastic band (NEB) calculations to determine the activation barriers for the first step to occur, i.e. to change a B-N bond to B-C and N-C bonds (Fig. 2(d)). Activation energy required is 1.63eV/atom suggesting that this process can happen only at elevated temperatures. At room temperature that is why the pristine $(BN)_m(C_2)_n$ should be stable and so are the phase separated ones.

There can be two different patterns for phase segregated BCN alloys. One is the phase separated island-like and the other one is a striped pattern. The island-like pattern consists of larger graphene-BN interface region than that in the striped pattern. This means that the number of B-C and N-C bonds are less in striped pattern than in an island form. As we have discussed earlier, the maximization of C-C and B-N bonds thus favors a striped pattern [13].

Till now, we have discussed mainly the stability issues of $(BN)_m(C_2)_n$ with 1:1 ratio and phase separated BCN alloys. A distributed mixture of BN and graphene with different $m : n$ ratios can also form depending on the growth condition. Different isomeric structures are also possible for a particular $m : n$ ratio. In the following section we are going to present a DFT study to analyze the stability and electronic properties of $(BN)_m(C_2)_n$ with different $m : n$ ratios. Utilizing the concept of aromaticity, the aim is to find out stable isomers for a particular $m : n$ ratio and also to explore the possibilities of achieving desired electronic properties.

Aromaticity, as extensively used to determine the stability of organic molecules, can provide us a working principle for determining stability of the structures as well. Benzene (C_6H_6) is the prototype for the organic molecules, which are stabilized by aromaticity. Borazine $(B_3N_3H_6)$, an isoelectric BN analogue of benzene on the other hand has one-third stability of benzene from the point of view of aromaticity [18–21]. This is particularly interesting in $(BN)_m(C_2)_n$, as the admixture of two not only changes the electronic property but also affects its stability. To investigate a stable isomer, our first working principle thus is to maximize the carbon hexagons, which essentially mimic benzene rings. A carbon-hexagon again can be surrounded by BN and each hexagon can be kept aloof or all hexagons can form a carbon-pathway. In a carbon pathway, π-conjugation is allowed whereas it is hindered in isolated C-hexagons.

To look for reasonable isomers, we consider that the following structural possibilities will not occur. Firstly, a hexagon will not contain B and N in 1 and 3 positions with respect to each other. These kind of structures are described by zwitterionic and biradical resonance structures, which basically result in an odd number of π-electrons on two of the Cs in the hexagon (Fig. 3).

Hence, a B-N pair should be placed either in 1,4 or 1,2 position in the hexagon with respect to each other. π-electrons will thus be distributed over a C-C bond and form a resonance structure. Second kind of structural constraint, that we consider, is the absence of B-B or N-N bonds. As discussed earlier, these kind of bonds result in the lowering of π-bonds and thus decreased relative stability of an isomer.

The relative positions of B and N around an all C hexagon is also a key factor that controls the electronic properties. To illustrate the phenomenon, let's consider the following two isomers. As in Fig. 3, the isomer I and isomer II, both have similar chemical configuration. But in Isomer I, B and N are connected to C at position 1 and 4 in the hexagon, which we can call B-ring-N para-arrangement. A donor-acceptor (D-A) interaction is thus established in this kind of structural arrangement. On the other hand in isomer II, B and N are connected to 1st & 2nd (4th & 5th) positioned C atoms in the hexagon. Although a D-A interaction occurs between neighboring B and N, B-ring-N interaction is forbidden. The local D-A interaction around a C-hexagon, as shown in Fig. 4, increases the HOMO-LUMO gap whereas N-ring-N (or B-ring-B) para arrangement results in the lowering of the HOMO-LUMO gap.

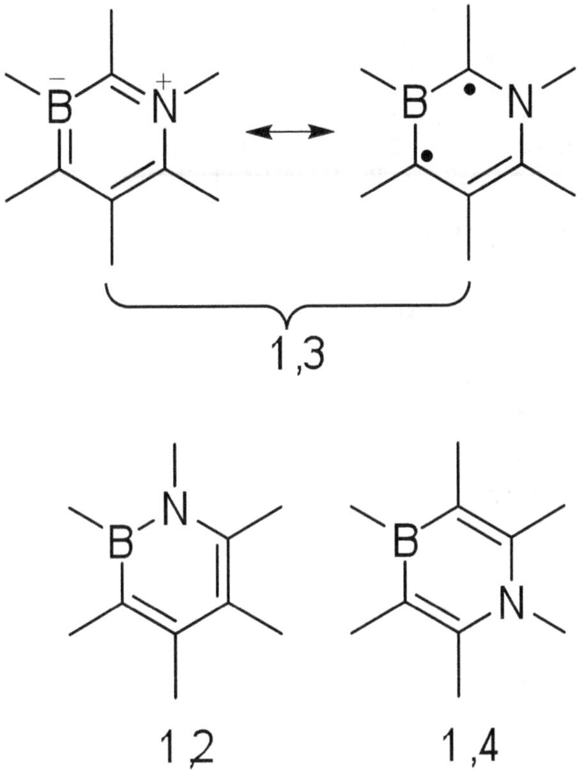

Figure 3. Schematic representation of zwitterionic and biradical resonance structures. Reprinted with permission from J. Phys. Chem. C 115, 10264 (2011). Copyright (2011) American Chemical Society.

We have performed density functional calculations to investigate the isomers of $(BN)_m(C_2)_n$ [22]. All the structures are optimized with both (Perdew-Burke-Ernzerhof) PBE [23] and (Heyd-Scuseria-Ernzerhof) HSE [24] functionals. The functionals based on local spin density approximation or generalized gradient approximations reproduce the structural parameters reasonably well, whereas the band gaps come out to be much smaller compared to experiments. The reason behind this is the self interaction error. HSE, with a better description of exchange and correlation within hybrid DFT, yields a band gap, which is much closer to the experimental value.

The degree of aromaticity is calculated quantitatively, with a harmonic oscillator model of aromaticity (HOMA) prescribed by Krygowski et al.[25]. The HOMA value of an ideal aromatic compound (Benzene) will be 1, whereas the value will be close to zero for non aromatic compounds. Anti-aromatic compound with the least stability will have a negative HOMA value. As mentioned earlier, the aim is to find the important isomers with relatively high stability and reasonable band gaps among $(BN)_m(C_2)_n$ compounds with $m : n$ ratios 1:1, 2:1, 1:3 and 2:3. Let's focus on each type separately.

2.1. 1:1 h-BN:Graphene (BC$_2$N)

We have considered six isomers for BC$_2$N, among which two structures BC$_2$N-I and BC$_2$N-II consist of all C-hexagon pathways. In the third one, BC$_2$N-III, all C-hexagons are connected linearly as in polyacenes where as the fourth one , BC$_2$N-IV, has disconnected all-C-hexagons. The other two structures, BC$_2$N-V & BC$_2$N-VI do not have any all-C hexagon but BC$_2$N-VI has at least polyacetylene paths whereas BC$_2$N-V has only isolated C-C bonds. Although there

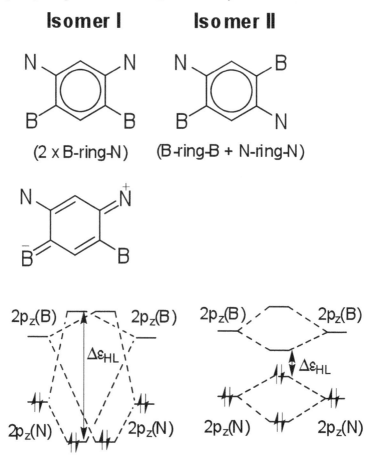

Figure 4. Qualitative representation of opening up a band gap and D-A interaction in isomer I and reduction of band gap in isomer II, with molecular orbital diagrams and valence bond representation. Reprinted with permission from J. Phys. Chem. C 115, 10264 (2011). Copyright (2011) American Chemical Society.

are several other isomers possible, we limit ourselves with these and try to understand the properties with the knowledge of aromaticity and conjugation. Firstly, the first three isomers, among all six are most stable and the relative energies differ by at most 0.15 eV (PBE) and 0.07 eV (HSE). The presence of all C-hexagons connected to each other not only increases the stable C-C and B-N bonds but also helps in the π-conjugation. The result is reflected in the HOMA values of first two structures, which are 0.842 and 0.888 respectively. This suggests the formation of aromatic benzene like all-C hexagons. The HOMA value of BC_2N-III is little less (0.642) but this structure in particular is not stable due to aromaticity rather due to the formation of polyacetylene paths. A slightly lower HOMA value observed in BC_2N-I compared to BC_2N-II is due to the difference in B-C bond (0.02Å), which leads to a change in D-A interaction.

If we look at the formation energies of BC_2N-IV & BC_2N-VI, the values are quite close. BC_2N-IV consists of completely isolated all-C hexagons. This is the reason of having high aromaticity of 0.88. But at the same time this increases N-C & B-C bonds and restricts π-conjugation. Therefore, this structure is less probable thermodynamically. BC_2N-VI, which was suggested to be the most stable BC_2N structure by Liu *et al.*[15], on the other hand has no aromatic all C-hexagon. But this structure contains all-C polyacetylene paths with C-C bond length 1.42 Å, which explains its low formation energy. BC_2N-V is the least stable among all, which has neither all-C hexagon nor polyacetylene C-paths. Obviously most unstable B-C and B-N bonds are maximized here creating an enormous strain in the structure. The presence of only C-C bond of 1.327 Å explains that. These factors make this compound thermodynamically most unstable among all five structures.

All these results give us a stand point from where we can judge the thermodynamic stability of other $(BN)_m(C_2)_n$ structures with the following working principles in hand:

(a) π-Conjugation within all C path increases stability.

(b) The formation of aromatic all C-hexagons also does the same, while this is more effective when hexagons are connected.

(c) There is not much contribution of B-ring-B or B-ring-N arrangement of B and N around poly(para-phenylene) (PPP) path in total energies. But indeed these, as discussed earlier, will affect the band gap, which we will present in the following section.

Coming to the band gap issue, the first three structures, which are close in energy, have band gaps ranging from 1.6 to 2.3 eV in HSE calculations (0.7 to 1.7 eV in PBE). The difference in BC_2N-I & BC_2N-II comes from the arrangement of B and N around all-C hexagon. As discussed in Fig. 3, D-A interaction increases for para-positions (i.e.1,4 or 2,6), which is observed in BC_2N-I. The band gap is 0.5 eV (0.65 eV in PBE), higher than that in BC_2N-II, where B and N are oriented in ortho-position (i.e.1,2 or 4,5). Quite obviously, BC_2N-III has all-C chain, which resembles a graphene nanoribbon and has the least value of the band gap.

2.2. 2:1 h-BN:Graphene (BCN)

We have investigated three structures of BCN, which have recently been synthesized [26]. The first one (BCN-I) has aromatic all-C hexagons connected in PPP path whereas the second one (BCN-III) contains all-C hexagon but connected in zigzag polyacene bonds. The final one (BCN-IV) consists of neither all-C hexagon nor a stripe of all-C region. BCN-I & BCN-II are iso-energetic, which is expected and is $\sim$ 0.5 eV lower than BCN-IV. This again explains the importance of aromatic all-C hexagon and π-conjugation. The absence of these and also the increased B-C, N-C bonds make BCN-IV relatively unstable. Another key point in BCN-I & BCN-I is the position of B and N around the hexagon. The stability may not be affected but the band gap is definitely changed by this. As expected, from the discussion in Fig. 4, BCN-I has a quite high value of the band gap. Aromaticity is higher in BCN-I (0.846) compared to BCN-III (0.557) , which is also seen in BC_2N structures. Iso-energetic BCN-I & BCN-II are equally probable during growth process but with a band gap range 1.3 to 2.7 eV .

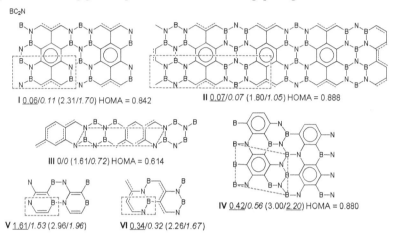

Figure 5. Six isomers of BC_2N are considered for this study. Relative energy (per formula unit) with respect to most stable structures and band gaps (in parentheses) are shown from HSE (normal print) and PBE (italic) calculations. HOMA values of all-C hexagons, obtained from HSE calculations are also provided. Reprinted with permission from J. Phys. Chem. C 115, 10264 (2011). Copyright (2011) American Chemical Society.

2.3. 2:3 & 1:3 h-BN:Graphene (BC_3N & BC_6N)

We now gradually increase the C-percentage with the anticipation of lowering the band gap because of increased graphene region. Two structures of BC_3N and three structures of BC_6N have been examined. Both the structures of BC_3N consist of aromatic hexagonal all-C rings connected in PPP path. The similarity in HOMA values depicts that picture. The position of B and N around C-ring is different though in BC_3N-I and BC_3N-II. The para arrangement of B-ring-N results in a large band gap in BC_3N-I (2.14 eV) while the ortho arrangement of the same in BC_3N-II lowers the value of the band gap. The stability is not affected by that fact as aromaticity and π-band formation are quite similar, which make those structures iso-energetic. A similar situation is also seen in BC_6N structures. BC_6N-I and BC_6N-II are iso-energetic mainly due to a similarity in structures. Both of them contain all-C rings connected in PPP path. But in the third one (BC_6N-III), all-C rings are separated, which makes this structure relatively unstable due to restricted π-conjugation. The HOMA value is maximum (0.93) in BC_6N-III, whereas the values are 0.78 and 0.86 respectively for BC_6N-I and BC_6N-II. The para arrangement of B-ring-N in BC_6N-I and BC_6N-III leads to large band gaps (1.58 and 1.34 eV respectively) while ortho-positioning of the same results in a reduced band gap in BC_6N-II. Finally, we have summarized the calculated (HSE and PBE) band gaps for all $(BN)_m(C_2)_n$ isomers (Fig. 7A and Fig. 7B). As expected, the band gaps are increased for HSE functional. Apart from that, both HSE and PBE -level calculations show a similar trend. A

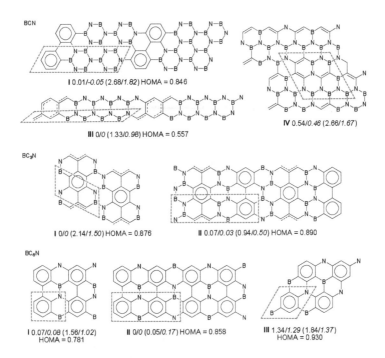

Figure 6. The isomers of BCN, BC$_3$N, BC$_6$N. Relative energy (per formula unit) with respect to most stable structures and band gaps (in parentheses) are shown from HSE (normal print) and PBE (italic) calculations. HOMA values of all-C hexagons, obtained from HSE calculations, are also provided. Reprinted with permission from J. Phys. Chem. C 115, 10264 (2011). Copyright (2011) American Chemical Society.

general trend is observed that an increase in the graphene region reduces the band gap. All the lowest energy structures for different compositions of $(BN)_m(C_2)_n$ have band gaps around 1 eV, which is a desired value for technological applications. One very important thing that we learnt from this study is that the position of B and N around C-ring controls the band gap without affecting the stability.

3. Functionalization

Incorporation of magnetism in 2D sp-materials has been an important point of discussion in recent times. The combination of localized moments of 3d transition metal atoms and the sp electrons of the host 2D lattice can give rise to interesting magnetic properties relevant to nano devices based on the principle of magnetoresistance, for example. Ferromagnetic long-ranged order, half metallicity, large magnetic anisotropy, electric field driven switching of magnetization etc. are being studied for transition metal atoms adsorbed on graphene and 2D BN sheets. Another important point is the adsorption of these species at the interface between BN and graphene. A recent study [27] based on first principles electronic structure calculations has revealed some interesting electronic and magnetic properties of Fe, Co and Ni adatoms adsorbed on a h-BC$_2$N sheet. A hexagonal site at the interface between BN and graphene

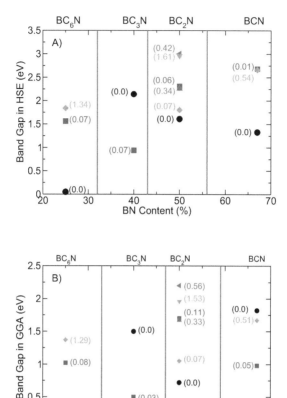

Figure 7. Band gaps of different isomers of $(BN)_m(C_2)_n$ are plotted against BN density, obtained from both (A) HSE and (B) PBE calculations. Relative energies are given in parentheses.

turns out to be a favorable site for adsorption. The presence of Fe, Co and Ni makes the system a magnetic semiconductor, a magnetic semi-metal and a non magnetic semiconductor respectively. Another interesting observation was that the adatoms are highly mobile on the surface and hence have the possibility to have a clustered configuration among themselves. It is interesting to note that the properties of this hybrid system are tunable in the sense that they can be modified by having different combinations of the width of each subsystem, BN and graphene.

Not only by magnetic adatoms, but by intrinsic edge properties, one can render magnetism in these 2D sheets. By DFT calculations, Dutta *et al.* [28] predicted interesting magnetic properties of H-passivated zigzag nanoribbons (ZGNRs) of various widths, doped by boron and nitrogen, keeping the whole system isoelectronic with C atoms in graphene. In the extreme case, all C atoms of ZGNRs are replaced by B and N atoms and zigzag BN nanoribbons are formed. In the ground state, the two edges are antiferromagnetically coupled and remain so for all dopings. However, the application of an external electric field affects the electronic structure of the nanoribbon giving rise to semiconducting and half-metallic properties. Electric-field induced changes in the magnetic properties are very interesting from a technological point of view. Other related studies [29] based on DFT revealed energetics, electronic structure and magnetism of quantum dots and nanorods of graphene embedded in BN sheet. It was showed that the formation energies and the HOMO-LUMO gaps of quantum dots vary as $1/\sqrt{n}$, where n is the number of carbon atoms in the dots.

Adsorption of gases on 2D materials is an important topic from the technological and environmental points of view. Materials for clean energy are always sought for and in this respect, efficient hydrogen uptake of suitable materials is an important issue. Raidongia et al. [26] have studied H_2 adsorption on BCN at 77 K and 1 atm. pressure. From their experiments, H_2 uptake of 2.6 wt % was observed. Also, CO_2 adsorption is very important for environmental issues. In their study, BCN was found to have a very high CO_2 uptake of 100 wt % at 195 K and 1 atm. pressure. It should be noted that the uptake is only 58 % by the activated charcoal under identical conditions. At room temperature and 40 bar pressure, the CO_2 uptake was found to be 44 wt %.

In a recent theoretical study, Cao et al. [30] showed that a zigzag interface between BN and graphene can have a strong capability of adsorbing hydrogen, much stronger than pure graphene, BN or the armchair interface between them. Moreover, the adsorption of hydrogen induces a semiconductor to metal transition. As the mobility of hydrogen on the surface is rather high, the hydrogen atoms can migrate to the zigzag interface and hence will increase the density of hydrogen storage with the added functionality of band gap engineering.

4. Summary and outlook

Graphene-BN nanocomposites offer a huge potential in various technological sectors, e.g., nano electronics, gas sensing, hydrogen storage, nanomagnetic storage devices, to name a few. The unique combination of these two materials with different electronic properties, forming a 2D network, offers many possibilities for studying fundamental science and applications for nanotechnology. However, many challenges, both in the domains of experiment and theory, will come in the way. Experimental synthesis of samples of good quality and state-of-the-art characterization techniques to reveal the atomic scale physics will be the issues. From the point of view of theory, one faces difficulties in having a correct description of the band gaps and electronic structures in standard approximations of materials-specific theories. However, with the availability of powerful supercomputing facilities, it is nowadays possible to treat large systems by sophisticated many body theories to have a much better quantitative descriptions. Nevertheless, one may envisage many interesting directions for the applications of these nanocomposites to utilize the interface properties of BN and graphene. One of them is the spin switching properties of organometallics adsorbed at the interface, similar to what has been studied recently [31] for a 2D graphene sheet. The other application can be the adsorption of amino acids [32] at the interface to increase the activity by their immobilization. Hopefully, in near future, we will observe many applications of these nanocomposites, useful for the human society.

Author details

Sumanta Bhandary and Biplab Sanyal
Department of Physics and Astronomy, Uppsala University, Box 516, 751 20 Uppsala, Sweden

5. References

[1] K. S. Novoselov, A. K. Geim, S. V. Morozov, D. Jiang, Y. Zhang, S. V. Dubonos, I. V. Grigorieva, A. A. Firsov, *Science* 306, 666 (2004).

[2] L. D. Landau, Phys. Z. Sowjetunion 11, 26 (1937).

[3] R. E. Peierls, Ann. I. H. Poincare 5, 177 (1935).

[4] A.K. Geim and K.S. Novoselov, Nat. Mater. 6, 183 (2007); A.H. Castro Neto *et al.*, Rev. Mod. Phys. 81, 109 (2009); A.K. Geim, Science 324, 1530 (2009).

[5] M. I. Katsnelson, Materials Today 10, 20 (2007).

[6] P. R. Wallace, Phys. Rev. 71 622-634 (1947).

[7] J. O. Sofo, A. S. Chaudhari and G. D. Barber, Phys. Rev. B 75, 153401 (2007); D. C. Elias *et al.*, 323, 610 (2009); O. Leenaerts, H. Peelaers, A. D. HernǦndez-Nieves, B. Partoens, and F. M. Peeters, Phys. Rev. B 82, 195436 (2010).

[8] M. Klintenberg, S. Lebegue, M. I. Katsnelson, and O. Eriksson, Phys. Rev. B 81, 085433(2010).

[9] M. Y. Han, B. Özyilmaz, Y. Zhang, and P. Kim, Phys. Rev. Lett. 98, 206805 (2007).

[10] Y.-W. Son, M. L. Cohen and S. G. Louie, Nature 444, 347 (2006).

[11] S. Bhandary, O. Eriksson, B. Sanyal and M I. Katsnelson, Phys. Rev. B 82,165405 (2010).

[12] P. Hohenberg and W. Kohn, Phys. Rev. B 136, 864 (1964); W. Kohn and L. J. Sham, Phys. Rev. A 140, 1133 (1965).

[13] H. Nozaki and S. Itoh, J. Phys. Chem. Solids, 57, 41 (1996).

[14] L. Ci, L. Song, C. Jin, D. Jariwala, D. Wu, Y. Li, A. Srivastava, Z. F. Wang, K. Storr, L. Balicas, F. Liu and P. M. Ajayan, Nat. Mater. 9, 430 (2010).

[15] A. Y. Liu, R. M. Wentzcovitch and M. L. Cohen, Phys. Rev. B 39,1760 (1989).

[16] K.-T. Lam, Y. Lu, Y. P. Feng and G. Liang, Appl. Phys. Lett. 98, 022101 (2011).

[17] K. Yuge, Phys. Rev. B 79, 144109 (2009).

[18] P. v. R. Schleyer, H. Jiao, Pure Appl. Chem 68, 209 (1996).

[19] E. D. Jemmis and B. Kiran, Inorg. Chem. 37, 2110 (1998).

[20] P. W. Fowler and E. J. Steiner, J. Phys. Chem. A 101, 1409 (1997).

[21] W. H. Fink, J. C. Richards, J. Am. Chem. Soc. 113, 3393 (1991).

[22] J. Zhu, S. Bhandary, B. Sanyal and H. Ottosson, J. Phys. Chem. C 115, 10264 (2011).

[23] J. P. Perdew and Y. Wang, Phys. Rev. B 45, 13244 (1992).

[24] J. Heyd and G. E. Scuseria, J. Chem. Phys. 120, 7274 (2004); J. Heyd, G. E. Scuseria and M. Ernzerhof, J. Chem. Phys. 118, 8207 (2003).

[25] T. M. Krygowsky, J. Chem. Inf. Comput. Sci. 33, 70 (1993).

[26] K. Raidongia, A. Nag, K. P. S. S. Hembram, U. V. Waghmare, R. Datta and C. N. R. Rao, Chem. Eur. J. 16, 149 (2010).

[27] P. Srivastava, M. Deshpande and P. Sen, Phys. Chem. Chem. Phys. 13, 21593 (2011).

[28] S. Dutta, A. K. Manna and S. K. Pati, Phys. Rev. Lett. 102, 096601 (2009).

[29] S. Bhowmick, A. K. Singh and B. I. Yakobson, J. Phys. Chem. C 115, 9889 (2011).

[30] T. Cao, J. Feng and E. G. Wang, Phys. Rev. B 84, 205447 (2011).

[31] S. Bhandary, S. Ghosh, H. Herper, H. Wende, O. Eriksson and B. Sanyal, Phys. Rev. Lett. 107, 257202 (2011).

[32] S. Mukhopadhyay, R. H. Scheicher, R. Pandey and S. P. Karna, J. Phys. Chem. Lett. 2, 2442 (2011).

Chapter 3

Properties of MWNT-Containing Polymer Composite Materials Depending on Their Structure

Ilya Mazov, Vladimir Kuznetsov, Anatoly Romanenko and Valentin Suslyaev

Additional information is available at the end of the chapter

1. Introduction

Carbon nanotubes (CNTs) are tubular structures composed of curved graphene sheets with diameter up to several tens of nanometers with typical length up to several micrometers. Single- and doublewall CNTs have diameters from 1.2 to ca. 3 nm and are usually packed in relatively dense structures ("ropes"). Multiwall carbon nanotubes can contain up to tens of concentrically aligned tubules and have diameter from 3-4 to tens of nanometers. Carbon nanotubes, both single- and multiwall, show outstanding mechanical and electrical properties [1, 2]. Nowadays CNTs are regarded as one of the key materials for development of various nanotechnology applications – new materials, sensors, actuators, field emitters etc. [3, 4, 5, 6]. In the last decade great effort was done in this field by many research groups, investigating structural, physical, mechanical, and electrical properties of CNTs.

Among all types of nanotubes single-wall nanotubes (SWNTs) were widely recognized as most perspective in regard of their predicted properties. Depending on chirality and diameter SWNTs can show significantly different electronic structure thus revealing metallic of semiconducting properties [7, 8]. Mechanical properties of SWNTs were investigated both theoretically and experimentally and were shown to outstand all other construction materials such as steel, carbon fibers etc [9].

Extremely remarkable properties of SWNTs are strongly limited in usage by their high cost and low yield of production methods. Commonly used methods of SWNT synthesis include arc discharge [10], laser evaporation of carbon targets [11], or catalytic decomposition of gaseous carbonaceous species (carbon monoxide [12], alcohols [13], various hydrocarbons [14]) by CVD technique [15]. As-produced SWNTs need purification from amorphous carbon and other graphene-like species (fullerenes, multiwall CNTs etc.) which is usually performed by strong oxidative media and/or by selective surfactants (followed by ultra-centrifugation etc.). Involvement of such complex techniques results in high cost of resulting material, especially in the case of production of CNTs with tailored mechanical and electronic properties. According to market analysis the average price of highly purified SWNTs (90-99 wt. %) lays in range 200-600 $ per gram depending on purity, chirality and surface composition. Note that high amounts of SWNTs are still less available.

Multiwall carbon nanotubes (MWNTs) were firstly described in 1953 and now are one of the most common and widely used nanotubes allotrope. Multiwall nanotubes are composed with several concentrically aligned tubular graphene sheets, with typical diameter in range 8-30 nm. Physical and mechanical properties of MWNTs are significantly lower than that for SWNTs but still are higher than properties of commonly used construction materials and reinforcement additives.

Multiwall carbon nanotubes can be synthesized in the same way as SWNTs by arc discharge, graphite evaporation by laser irradiation or by catalytic decomposition of gaseous carbon-containing species [16] by CVD. The last one is the most perspective due to possibility to regulate CNT diameter and length; due to high yield and high selectivity of the process less or even no purification by aggressive oxidation is needed to achieve MWNT with high purity (higher than 90 wt. %) [17]. CVD process has high scaling potential, e.g. by realization of the fluidized bed technique [18, 19, 20].

In the last few years significant progress was achieved in the scaling of synthesis of MWNTs by catalytic CVD route. Several companies have demonstrated large-scale facilities for the process, for example, Bayer AG (Germany), Nanocyl (Belgium), Arkema (France), Hyperion Catalyst (USA), CheapTubes Inc. (China). Development of the large-scale synthesis route for MWNTs with high purity and relatively low defectiveness allowed to significantly lower market price for such product which is in range 1-15 $ per one gram depending on purity, mean diameter and surface functionalization.

Relatively high availability of MWNTs and their remarkable properties result in great interest of their usage in various nanotechnology applications. At the present time high amount of research work was done in the field of MWNT investigation and application. Multiwall carbon nanotubes can be used as components of composite materials with polymer, metal or ceramic matrices [21]; as chemical sensors [22, 23]; as components of catalytic systems [24, 25]; as electromagnetic shielding materials [26]; for biomedical applications such as selective drug delivery [27, 28] etc.

One of the most perspective approaches of usage of MWNT's superior properties is development of new multi-functional composite materials with improved and tailorable properties. Such composites can be used in various applications, for example as construction materials [29], anti-static coatings [30], low-weight electromagnetic shielding [31], conductive polymers [32] etc. Polymer matrices are mostly used for development of such composite due to their light weight, low price, good processability and controllable chemical, physical and mechanical properties as well as good scaling perspectives.

To date several tens of polymer matrices were investigated for the synthesis of MWNT-and SWNT-loaded polymeric composites. These are epoxy resins [33], polyurethanes [34], polyolefines [35, 36], polymethylmethacrylate [37], polystyrene [31], and others.

Systematic investigation of properties of novel multifunctional composite materials, containing carbon nanotubes is of essential importance to understand and improve their properties.

It is known [24, 38] that CNTs of same type but produced by different vendors often show significant difference in chemical and physical properties depending on their diameter, length, defectiveness, agglomeration state, surface chemistry etc. Variation in properties of the filler may result in non-linear changes in properties of resulting composite. Thus investigation of properties of CNT-containing composites depending on properties of incorporated nanotubes and polymer host matrix is an important task.

In this chapter we describe an attempt to systematic investigation of structural, physical-chemical, and electrophysical and electromagnetic properties of thermoplastic composite materials, comprising multiwall carbon nanotubes with different mean diameters and morphology.

2. MWNT-containing polymer composites

2.1. Approaches to synthesis, main problems

As it was mentioned above, CNT (and, first of all, MWNT)-containing polymer-based composite materials attract great interest in the last decade. Great work was done in this field and tens of various host matrices were investigated. Several main problems in the area of design and preparation of nanotube-filled composite materials can be outlined according to literature analysis.

1. As-synthesized carbon nanotubes are usually arranged either in dense aligned arrays ("ropes") or tangled "furballs", composed of several tens of closely matted CNTs. First type is more typical for SWNTs, and the second one is most typical for CVD-produced MWNTs. Dense entangled MWNT arrays should be destructed during composite synthesis in order to achieve maximum dispersion state of nanotubes and subsequent maximum increase in properties of the composite. Also in this case it is possible to reach electrical percolation threshold at relatively low concentration of the filler due to intensive linking of high-dispersed nanotubes between each other. See review [39] for details of dispersion of MWNTs in various liquids using different technique.

2. Carbon nanotube fillers in the polymer matrix (and, in common sense, in all types of composite materials) can act as reinforcement material in several ways.

The first, incorporated nanotubes with high mechanical properties and high electrical conductivity, may act in the same way as macroscopic fillers (carbon fiber, glass wool etc.) providing stress transfer from the low-strength matrix in the case of mechanical load or charge transfer through continuous linked conductive network in the case of electrical load [40, 41].

The second, CNT filler is providing nucleation sites for the growth of polymer nanosized crystallites. Introduction of high amount of nanotubes with high surface chemical potential results in significant reduction of the grain size of resulting composite [42, 43] thus leading to increase of its mechanical properties.

Thus it is of crucial importance to obtain high dispersion degree of nanotube filler in the bulk volume of the polymer matrix and provide intensive interaction between CNTs and polymer. These two tasks should be resolved during composite synthesis in order to obtain material with increased properties. This can be done in several ways depending on the type of the polymer used as matrix material.

All polymer materials can be roughly divided in two parts – thermoplastic and thermoreactive polymers. Both types were used for synthesis of MWNT-containing composites with certain success.

Thermoreactive matrices, such as epoxy resins and polyurethanes, were one of the first used for preparation of MWNT-loaded composites. The main route of preparation of such type of materials involves mixing of carbon filler with resin or with chosen intermediate solvent (acetone, dimethylformamide etc. [44]) which is later mixed with the resin. The mixing process is often assisted with ultrasonic treatment which results in higher dispersion degree of the nanotube filler in the matrix [45]. Obtained mixture is molded to form necessary shape and cured. The technique described is quite experimentally simple and scalable. However, due to high viscosity of epoxy resin it is hard to disperse entangled nanotubes uniformly in whole volume of the polymer. In the case of usage of intermediate solvents such as acetone, the last must be evaporated before curing process. Carbon nanotubes tend to spontaneous agglomeration while staying in suspension, thus destroying achieved dispersive state and resulting in lowering of composite's properties as compared with theoretically predicted.

Several special procedures can improve the process, such as chemical surface functionalization [46, 47] or usage of short aligned CNT arrays as starting material [48, 49], allowing to obtain good distribution of CNTs with low electrical percolation threshold and increased mechanical properties.

Thermoplastic polymers can be used as matrix materials for design and synthesis of nanotube-filled composites. These matrices can be processed by variety of techniques, such as solution casting, extrusion, pressure molding, hot pressing etc. Processing methods can be divided in two main parts – *temperature*-assisted and *solution*-assisted technique.

Twin-screw extrusion, pressure molding, hot pressing, liquid casting and similar methods can be described as *temperature-assisted*. The main step of these techniques involves melting of mixture of the polymer matrix material and CNT filler with subsequent processing of the melt blend [35]. Such approaches are relatively cheap, scalable and experimentally simple, but still have several disadvantages.

Usually the polymer blend has high viscosity in molten state thus preventing disaggregation of entangled agglomerates of CNTs. Usage of ultrasonic treatment is this case is complicated by high temperatures and closed volume of experimental setup. Moreover, increase of CNT loading results in further sharp increase of viscosity of polymer-nanotube blend, preventing achieving high dispersion of incorporated nanotubes.

These problems can be partially solved by using of additional mixing procedure, such as mechanical activation of the solid polymer powder with CNTs, or using of high CNT shear flow mixers for the molten mixture providing high dispersion degree. However, these procedures can result in breaking of nanotubes, especially in the case of multiwall CNTs with defective walls, with corresponding CNT shortening and decrease in mechanical and electrophysical properties of the composite.

Solution-assisted technique of preparation of nanotube-filled polymer composites includes dissolving of polymer material in appropriate solvent, mixing of the resulting solution with CNTs and subsequent evaporation of the solvent with formation of the polymer-nanotube film [50]. High disaggregation state of CNTs can be achieved by using ultrasonication and/or high-intensive mixing [51] of the nanotube-polymer suspension due to reasonably low viscosity of the solution. Thin films can be produced by this technique, allowing one to design functionally grade materials, electrostatic coatings, polarizing films etc.

However, this technique has some disadvantages – it is hard to obtain massive samples by solution casting and also special precautions must be applied in order to avoid CNT agglomeration during drying of the composite film (for example, surface functionalization, shortening of nanotubes etc. [52, 53]).

There are some specific methods of synthesis of CNT-loaded composite materials, which cannot be ascribed to abovementioned types, for example *in situ* polymerization and coagulation precipitation techniques. The first one includes deposition of the catalyst on the surface of CNTs with further polymerization [54] or radical polymerization of the monomer (e.g. polystyrene, polymethylmethacrylate) in presence of carbon nanotubes or other nano-sized fillers or monomers [37 ,55, 56, 57]. High dispersion degree of nanotubes can be achieved by such methods, with following processing using conventional techniques, mentioned above.

The second method of CNT/polymer composite synthesis, coagulation precipitation, was firstly developed by Du et al. [58] for SWNT/PMMA composites. The coagulation precipitation (CP) technique involves dissolution of the polymer in appropriate solvent, mixing of this solution with carbon nanotubes (or other filler materials). The resulting slurry is mixed with the second solution in which the first solvent is soluble, but the polymer is not. As a result the polymer/CNT mixture immediately precipitates, forming disperse composite material which can be later processed in usual ways.

Coagulation precipitation technique has several remarkable advantages as compared with abovementioned methods. By right choice of the first solvent it is possible to obtain both dissolution of the polymer and good wetting of nanotubes. For example, in the case of MWNT/PMMA or MWNT/PS composites dimethylformamide of N-methylpyrrolidone can be used for this task. These solvents are known to provide very stable CNT suspensions [59] and can dissolve corresponding polymers in high concentrations. The second solvent for this system is water.

High CNT dispersion state is produced by ultrasonic treatment of the polymer-solvent mixture and is quite stable during minutes. It is of crucial importance that such high dispersion can be "frozen" on the second step by mixing with the second solvent, thus there is no reason to obtain super-stable CNT suspension.

The precipitation of polymer starts immediately after mixing with the second solvent and this process proceeds at high rate, producing small particles. Dispersed carbon nanotubes act as nucleation sites allowing reaching intimate interaction between polymer and individual CNTs.

CP technique still has certain inconveniences – for example, lack of scalability potential. Solution pair should be chosen carefully, providing both mutual solubility, and partial solubility of the polymer matrix material.

In the present chapter we describe preparation of MWNT-containing composite materials using coagulation precipitation technique and polymethylmethacrylate (PMMA) and polystyrene (PS) as matrices. All these polymers are thermoplastics and can be processed using common pressing and extrusion techniques.

2.2. Experimental: Synthesis of MWNTs and MWNT-loaded composites

Multiwall carbon nanotubes were synthesized *in-lab* by ethylene decomposition over bimetallic FeCo catalyst in hot-wall CVD reactor at 680 °C. Details of preparation technique can be found elsewhere [60]. As-prepared MWNTs were additionally purified by reflux in HCl (15 wt. %) during 3 hours, washed with distilled water and dried in air at 80 °C for 24 hours.

Preparation of MWNT-loaded composites was performed using coagulation precipitation technique.

MWNT/PMMA and MWNT/PS composites were synthesized in similar way as follows. Polymer powder (PMMA, m.w. ~ 100000; Polystyrene, m.w. ~ 120000) was dissolved in dimethylformamide with concentration 0.1 g/ml. Calculated amount of air-dried MWNTs was loaded in water-cooled glass reactor and poured with polymer/DMF solution (40 ml) diluted with pure DMF (40 ml).

Resulting mixture was sonicated by Ti horn ultrasonicator (output power 8.5 W/cm^2) during 15 minutes under constant water cooling (the temperature of mixture was not higher than 50 °C). Resulting slurry was poured in ~ 1.5 liters of distilled water (t = 60-70 °C) under vigorous stirring immediately after US treatment. Precipitation of the polymer-MWNT composite proceeds immediately, resulting in formation of spongy-like deposit, which was left to stay overnight to complete coagulation.

The precipitate was filtered using Buchner funnel, washed with water (3×500 ml) and dried in air at 60 °C overnight. Residual water was removed by drying in vacuum (10^{-2} torr) at 60 °C for 2 hours.

Composite powder can be processed using common technique. In our case hot pressing was chosen as one of the most simple ways to make polymer films. Powder was placed between two polished steel plates, which were heated up to melting temperature of the polymer and pressed with hydraulic press with pressure ca. 400 kg/cm^2. Copper ring with 0.5 mm thickness was used as spaces. Produced composite films were Ø 60×0.5 mm^3 in dimensions.

Scanning and transmission electron microscopy was used for investigation of the structure of composite powder and films (JSM6460LV and JEM 2010 electron microscopes were used). Powder samples were placed on the conductive carbon adhesive tape, films were cut in plates with size ca. 8×3×0.5 mm^3, which were broken and glued to the copper stand with breaks upwards using silver glue. In order to avoid surface charging during SEM investigations all samples were additionally covered with 5-10 nm gold layer.

Electrophysical properties of composite films were investigated using four-probe technique with silver wires connected to sample surface with silver glue.

Electromagnetic response properties of free-standing polymer composite films were investigated in frequency range 0.01-12 GHz, 26-37 GHz. Array of experimental setups was used: quasi-optical setup based on panoramic meter KSvN R2-65 (Russia), Mach-Zehnder interferometer based on backward wave oscillator (Russia), HP Agilent PNA 8363B network analyzer (Agilent, USA) with multimode resonators (for dielectric permittivity spectra), and R2M-04 reflectance/transmission meter (Mikran, Russia). For ε spectra measurement samples were cut in pieces with size 0.5×2.5×30 mm^3 and placed in antinode of electric field parallel to electric field lines. Transmission coefficient was measured using polyfoam gasket, samples were cut in rings with an outer diameter Ø16.5 mm and inner diameter Ø6.95 mm and glued to the gasket. The measuring setup was calibrated using pure polyfoam gasket covered with glue.

2.3. Properties of pure MWNTs

Macro-scale properties of MWNTs and MWNT-based composites depend strongly on their nanoscale parameters, such as diameter, particle size distribution, morphology of agglomerates. In this work we have investigated three types of MWNTs differencing in main diameter (and diameter distribution), defectiveness and morphology of agglomerates. On Figure 1 TEM data and diameter distribution for all types of MWNTs is shown.

Mean diameter of MWNT (as obtained by statistical analysis of TEM images) makes the value of ~7-9 nm for MWNT[8], ~10-12 nm for MWNT[12], and ~22-24 nm for MWNT[22]. From TEM images one can roughly evaluate defectiveness of MWNT (amount of amorphous species on the surface of CNT, opened walls etc.) which is increasing with decreasing of their mean diameter.

According to SEM data the morphology of MWNT secondary agglomerates varies, changing from rope-like structure for "thin" tubes (MWNT[8], MWNT[12]) to tangled furball-like structure for "thick" tubes (MWNT[22]). On figure 2 SEM images of typical MWNT agglomerates are presented.

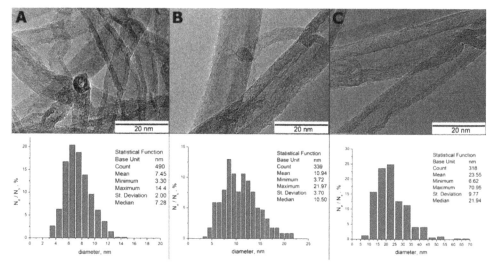

Figure 1. HRTEM images of MWNT samples used for preparation of MWNT/PMMA composites and their statistically calculated diameter distribution. A – MWNT[8], B – MWNT[12], C – MWNT[22].

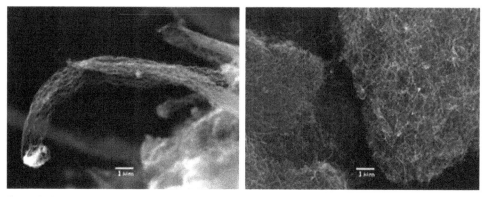

Figure 2. SEM micrographs of secondary aggregates of MWNTs. Left – MWNT[8], "rope-like" structure; right – MWNT[12], MWNT[22], "furball" structure.

Entangled nanotubes, which are forming "furball" structure, are hard to disperse in liquid media. MWNTs of all types tend to agglomerate in liquid dispersion, thus it is necessary to achieve certain conditions, allowing to "freeze" dispersed state of nanotubes. In our case usage of the coagulation precipitation technique allows to do this, as coagulation proceeds almost immediately after sonication and dispergation.

2.4. PMMA and PS composites – Influence of matrix type

Polymethylmethacrylate (PMMA) and polystyrene (PS) both are cheap, large-scale and easily processable polymer matrices, which are often used as model systems for synthesis of polymer composites. Polymers with similar average molecular weight were used for synthesis of composites, both PMMA and PS are soluble in DMF and precipitate using water as secondary solvent.

Principal physical-chemical properties of polymethylmethacrylate and polystyrene polymers are quite similar (see Table 1 for details) [61].

From the Table 1 it can be clearly seen that PMMA and PS have very similar physical and chemical properties and the main difference is their surface composition and polarity of the surface [62, 63]. Carboxylic functions are present in PMMA structure thus resulting in increased oxygen content and increased polarity of the surface. Polystyrene has no oxygen-containing groups, showing lower polarity and surface tension [64].

Polymer	Formula	Surface free energy	Surface oxygen (XPS)	Polarity[1]	Density, g/cm^3	Thermal conductivity, W/m×K	Dielectric constant, ε'@1MHz@25°C
PMMA		41.2 mJ/m^2 [65]	38.6 at. % [66]	0.28 [67]	1.17	0.19	2.80 [68]
PS		40.1 mJ/m^2	0 at. %	0.17	1.04	0.22	2.55

Table 1. Physical-chemical properties of polymethylmethacrylate (PMMA) and polystyrene (PS).

Thus investigation of these model polymer systems allows one to reveal basic principles of the influence of the matrix type on the properties of MWNT-loaded composites.

Transmission electron microscopy analysis of MWNT/PMMA and MWNT/PS composites was performed to investigate internal structure of materials. Corresponding micrographs are shown on figure 3. Interaction of carbon nanotubes with polymer matrix results in surface wetting of MWNTs with polymer, which depends strongly both on the surface composition of CNTs and on surface properties of the polymer. As-prepared nanotubes have low amount of polar (oxygen-containing) groups, thus surface is mostly hydrophobic. Chemical functionalization (e.g. oxidation) can modify surface composition of the materials, making it hydrophilic. According to our previous study [69], as-prepared CNTs have ca. 0.3-0.5 oxygen-containing groups per nm^2, so their surface is mostly hydrophobic, but still can show several hydrophilic behavior.

---

[1] Polarity calculated according to surface tension values, measured by sessile drop technique.

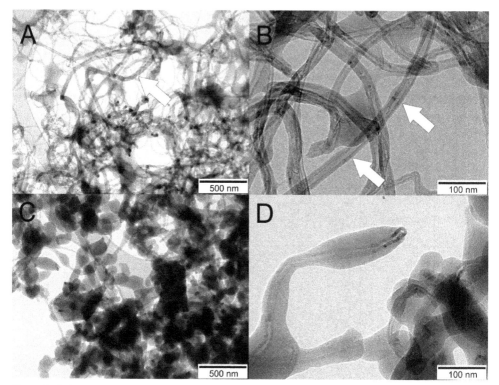

Figure 3. TEM images of MWNT-containing composites. A, B – MWNT[22]/PMMA (5 wt. %), C, D – MWNT[22]/PS (10 wt. %). Arrows are indicating naked parts of MWNTs (not covered with PMMA).

It is reasonable to suggest that hydrophilic surface would have higher affinity to polar polymer matrices, and pristine nanotubes would be better wetted by low-polar polymers.

Such behavior was observed for both compared MWNT/PMMA and MWNT/PS samples. As it can be seen from Figure 3, the surface of CNTs is covered with polymer. In the case of PMMA as a matrix, nanotubes are not fully covered with polymer. Both covered and naked surface areas are observed (marked with arrows on Fig. 3 A, B) for PMMA-treated composite. It is important to note that observed contact angles between nanotube surface and PMMA layer are still lower than 35°.

The possible reason for partial surface coverage with PMMA is certain irregularity of the functional composition of MWNT surface, for example, PMMA-covered areas are observed due to higher polarity of that nanotube part (e.g. due to partial oxidation), providing better wetting and lowering of the contact angle.

In contrast, PS-covered composite sample reveal significantly different structure where the surface of MWNTs is completely covered with thin continuous layer of the polymer (with thickness up to several nanometers). In the case of low CNT loading in composite one can observe disaggregated nanotubes, completely wrapped and wetted by PS (fig. 3, D). For higher nanotube content polystyrene "drops" are observed settled on the surface of nanotubes covered still with undestructured PS layer. As soon as polystyrene shows lower polarity as compared with PMMA and moreover has aromatic rings in structure, which have high affinity to the conjugated π-system of pristine (low-oxidized) nanotube, the wetting is more ease and coverage is higher in this case.

The contact angle between PS drops and CNT surface is in range 22-25° indicating high wetting ability of PS towards untreated surface of nanotubes.

Interfacial interaction between polymer matrix and incorporated nanotubes significantly affects structural and physical properties of composite films. Formation of dispersed nanotube array or network in the volume of the polymer matrix is essential to drastical change of material properties. Carbon nanotubes tend to agglomerate during synthesis of composites, thus decreasing dispersion degree.

Formation of CNT agglomerates in composite powder was observed for both PMMA- and PS-based materials using transmission electron microscopy. TEM micrographs of different nanotubes agglomerates are shown on Figure 4.

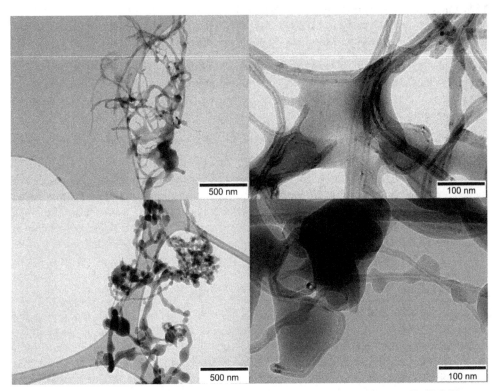

Figure 4. TEM images of agglomerated nanotubes in composite powders. A, B – MWNT[22]/PMMA (5 wt. %), C, D – MWNT[22]/PS (10 wt. %).

Two types of nanotube agglomerates can be distinguished in polymer-based composite materials. These agglomerates have been identified as "primary" and "secondary", according to their origin. *Primary agglomerates* are formed mainly during synthesis of carbon nanotubes themselves and can be avoided in growth-aligned nanotube arrays [70]. These aggregates must be destroyed during composite synthesis by ultrasonic treatment, high-shear flow mixing etc. Destruction of such particles in necessary to achieve high dispersion degree and, for example, correspondingly low percolation threshold for composites. Nevertheless, for CVD-grown nanotubes these aggregates are hard to destroy and they are still occurring in composites, covered with polymer layer. Such aggregates are typical for systems with low wetting of CNTs with polymer, e.g. for MWNT-PMMA (fig. 3-A & 4-A).

Agglomerates of secondary type are formed during the process of composite synthesis due to re-aggregation of dispersed nanotubes in solution due to van der Walls forces. In this case individual nanotubes are separated from each other in solution and linked by the polymer particle in solid composite (fig. 4-B&D). Strong wetting of carbon nanotubes with polymer assists destruction of primary agglomerates with subsequent formation of secondary-type agglomerates. Such phenomenon is observed for PS-based composites and is clearly seen on TEM micrographs – even for high MWNT loading (up to 10 wt. %) almost none primary agglomerates were observed. MWNTs are dispersed and separated from each other, forming secondary agglomerates, clued by polymer particles (fig. 4-C&D).

Investigation of composite films using scanning electron microscopy (SEM) was performed in order to elucidate influence of CNT-polymer interface on structural properties of composites. Corresponding SEM images of fresh breaks of composite films with different MWNT loading are shown on fig. 5.

For both types of polymer matrices separated nanotubes can be clearly seen on the surface of film breaks.

Difference in polymer-nanotube interface properties results in significant changes of the dispersion state of nanotubes in PMMA- and PS-based composites. For PMMA-based samples only single nanotubes can be seen for samples with 1 wt. % of nanotubes (fig. 5-A). Increase of MWNT content to 4 wt. % results in more uniform and dense dispersion of nanotubes in the volume of the polymer (fig. 5-C). Nevertheless, low-filled areas can be observed indicating relatively low dispersion degree of nanotubes. Certain amount of MWNTs can still occur in the agglomerated form, preventing formation of highly-disaggregated nanotube "network" in the polymer matrix.

Such phenomenon can be observed for the highest investigated MWNT loading in the composite (10 wt. %, fig. 5-E). Dense and uniform nanotube array can be clearly seen on the surface of composite film break. Note that the length of MWNT residues on the surface of the film is 1-2 μm, moreover, the surface of nanotubes is not covered with the polymer. This may be attributed to stretching of CNTs from the volume of the polymer during breakage of the film, which can be easily assumed taking into consideration low wetting ability of PMMA towards hydrophobic surface of untreated carbon nanotubes.

Polystyrene-based composite materials show opposite phenomenon of high wetting and high dispersion degree even at low MWNT loadings. Nanotubes in all MWNT/PS composite film samples are well-dispersed and covered with polymer layer. Uniform CNT distribution can be observed for samples with both low and high nanotube loading. According to SEM data nanotubes are randomly and evenly distributed on whole breakage area, which can be clearly seen on fig. 5-B, D, F.

Note that protruding parts of MWNT for PS-based composites are shorter as compared with similar PMMA-based samples and are wrapped with polymer layer. Such phenomenon can be attributed to higher wetting of CNTs with polystyrene and higher adhesion of filler to the material of the polymer matrix.

Thereby according to TEM and SEM data significant difference in wetting ability of investigated polymers towards MWNTs results in drastic changes of structural properties of composite powders and films. Highly-wetting aromatic polystyrene matrix allows one to reach higher dispersion degree at lower filler concentrations as compared with more polar polymethylmethacrylate due to mainly hydrophobic surface character of untreated nanotubes.

Area of application of novel materials is strongly dependent on their physical properties. As it was shown above, MWNT-based composites are perspective as tailorable materials for electrical and electromagnetic applications. Structure and dispersion state of nanotubes in composite, as well as their interconnections (i.e. formation of connected array of CNTs) strongly affect physical properties of composite materials. It is well known that introduction of continuous carbon nanotubes in the dielectric polymer matrix allows to increase conductivity of the resulting composite by several orders of magnitude. Conductive composite materials can be characterized by the percolation threshold which for multiwall carbon nanotubes lays in range from ca. 0.005 wt. % [71] to 3-4 wt. % [72] depending on the electrophysical properties of initial nanotubes and peculiarities of the composites' preparation (alignment and dispersion of nanotubes etc.).

Electrical conductivity measurements reveal significant difference between PMMA- and PS-based composites with similar MWNT loading. Conductivity data for PMMA-based composites with different types of MWNTs is shown on figure 6.

Electrical percolation threshold for PMMA-based composites was estimated as ~ 1-2 wt. %, reaching maximum value of 10^{-1}-10^{-2} S/m for 3-5 wt. % of CNT content.

PS-based composite materials show surprisingly low electrical conductivity. Significant conductivity (~ 4.5×10^{-4} S/cm) was observed for composite sample with 10.0 wt. % loading of CNTs, all other samples show electrical resistivity higher than 10^9 Ohm/cm which was the sensitivity threshold for the setup used for measurements.

As it was shown by electron microscopy investigations, PS-based composites reveal higher wetting and higher dispersion degree of nanotubes in the volume of the polymer as compared with PMMA-based materials. Unusually high percolation threshold and low conductivity value can be explained taking into consideration insulation of conductive nanotubes with wrapped polystyrene layer. In this case perfectly dispersed three-dimensional nanotube network in the bulk volume of the polymer is formed by *insulating* objects, preventing current flow through composite even at high CNT loadings.

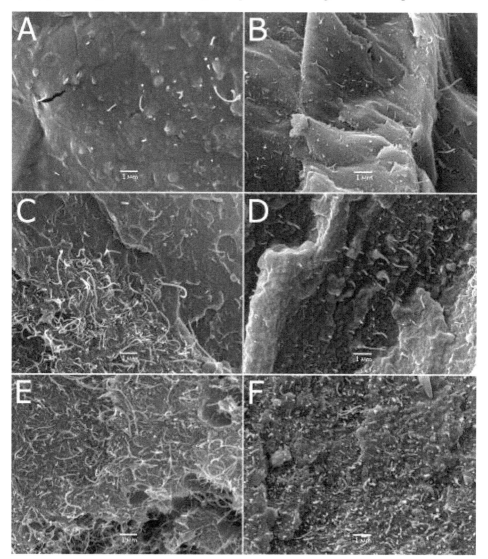

Figure 5. SEM images of MWNT/PMMA (left column) and MWNT/PS (right column) composite films. A, B – 1 wt. %, C, D – 4 wt. %, E, F – 10 wt. %.

In the case of poorly-wetted PMMA-based composites the surface of nanotubes is not completely insulated, thus allowing reaching of electrical saturation at relatively low CNT loadings. Moreover, destruction of primary agglomerates of nanotubes, observed for PS and not for PMMA-based samples, results in diminishing of electrical contacts between nanotubes due to their insulation. Residual CNT agglomerates in the case of PMMA-based composites facilitate formation of conductive paths in the volume of the polymer matrix. As a result overall electrical conductivity of MWNT/PMMA materials is higher as compared with PS-based samples.

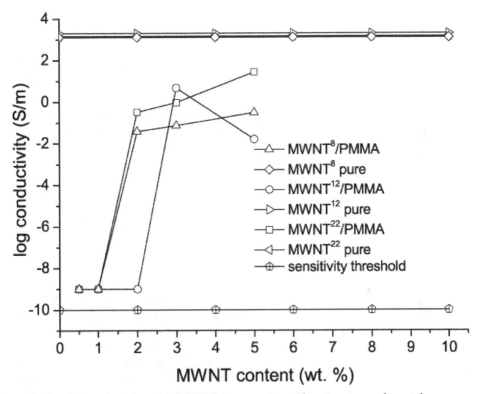

Figure 6. Electrical conductivity of MWNT/PMMA composites with various types of nanotubes versus filler concentration.

Electromagnetic response properties of composite materials play significant role in their application area. Incorporation of conductive media, such as carbon nanotubes, affects strongly the way of interaction of certain material with electromagnetic irradiation (EMI). Conjugated π-system, occurring in carbon nanotubes, allows both to dissipate and reflect electromagnetic wave, thus providing two possible mechanisms of EM response – reflectance and absorbance.

Electromagnetic response properties of MWNT-containing composites were investigated in broadband region (2-36 GHz) and was found to be strongly dependent on MWNT type and diameter as well as on polymer matrix type.

Influence of structural properties of MWNTs on EM shielding properties of composites was investigated for MWNT/PMMA materials due to their high electrical conductivity. Structure of initial CNT affects electrical and electromagnetic properties of CNT-based composites. We have investigated EM response properties of MWNT-based composites in frequency range 3-11 GHz (complex dielectric permittivity for MWNT/PMMA samples). Transmission (T) and reflection (R) coefficients can be easily calculated using following equations. Absorption coefficient of the EM radiation in the sample can be calculated as $A = 1 - R - T$.

$$R = \left| \rho \frac{1 - e^{-i2kd}}{1 - \rho^2 e^{-i2kd}} \right|; \quad T = \left| \frac{(1 - \rho^2) e^{-ikd}}{1 - \rho^2 e^{-i2kd}} \right|,$$

where $\rho = \dfrac{Z-1}{Z+1}$, $Z = \sqrt{\dfrac{\mu^*}{\varepsilon^*}}$ is wave impedance, and $k = \dfrac{2\pi f \sqrt{\varepsilon^* \mu^*}}{c}$ is the wavenumber, f is the frequency of the EM wave, c is the speed of light, d is the thickness of the sample, $\varepsilon^* = \varepsilon' - i\varepsilon''$ and $\mu^* = \mu' - i\mu''$ are complex permittivity and magnetic permeability of the investigated material correspondingly. Dielectric losses (ε' and ε'') are measured experimentally, and $\mu' = 1$, $\mu'' = 0$ (no magnetic losses are observed in the sample). Observed values of ε' and ε'' are growing with increase of the CNT loading in the composite and lay in range 20-70 for MWNT[22]/PMMA and 5-40 for MWNT[12] and MWNT[8]/PMMA composites.

On the figure 7 data on measured transmission and reflections coefficients are shown.

Increase of the conductivity of the composites with MWNT loading in all samples leads to growth of R and diminishing of T parameters. For example for the MWNT[8]/PMMA composite one can see higher R and lower T values for the sample with 3 wt.% of MWNT whereas for the sample with 5 wt. % these values are lower and higher, correspondingly, correlating with its conductivity. It should be mentioned that even composites with the lowest electrical conductivity (with CNT loading 0.5-2 wt.%) show high values of the R, which may be due to formation of isolated conductive MWNT structures in the volume of the polymer that cannot be registered by macroscopic measurements of the electrical conductivity but still can interact with electromagnetic field. Polarization of such isolated structures gives contribution in EM response of non-conductive materials with subsequent growth of the permittivity.

Note that transmission and reflection coefficients do not depend directly on the electrical conductivity for different MWNT types. The highest values of R are observed for the samples MWNT[22]/PMMA which show the lowest conductivity among other composites. This phenomenon may be described taking into account significant difference between number of individual nanotubes of each type incorporated into polymer matrix. Relationship between CNT number can be roughly estimated as relationship between $r_{oj}^2 - r_{ij}^2$, where r_{oj} is the outer diameter of CNT type j, r_{ij} is the inner diameter of CNT type j

Simple math gives rough approximate of relationship between CNT number in the composite with the same weight loading as $N(MWNT^{22}):N(MWNT^{12}):N(MWNT^8) \approx 8{:}4{:}1$.

Thus it is possible to propose that despite of the lower macroscopic electrical conductivity, composite materials comprising MWNT with lower diameter possess higher amount of polarizable species, giving higher values of dielectric permittivity and reflection coefficient.

All samples with low CNT content (0.5 wt. %) are almost transparent in all frequency range. Increase of MWNT loading leads to reduction of transmission coefficient. Transmission is strongly affected by MWNT type and is changing unidirectionally with the dependence of permittivity – the lowest transmission is observed for MWNT[22]-containing composites which have higher ε value as compared with MWNT[12]/PMMA, MWNT[8]/PMMA and composites.

For both matrices percolation-like concentration dependence can be observed with sharp decrease of EM transmission coefficient (and corresponding increase of EM shielding of composites) after 1-2 wt. % of MWNTs in materials.

Nevertheless, the most significant difference for investigated matrices is that in case of PS matrix EM shielding is provided mainly by absorption, growing with increase of MWNT loading, and for the case of PMMA matrix main part of EM shielding is constituted by EM reflectance. Principal changes in mechanisms of EM attenuation in polymer composites filled with same type MWNTs can be explained taking into consideration abovementioned differences in their structural and electrophysical properties.

PMMA-based composites show high electrical conductivity and relatively low percolation threshold therefore above percolation threshold such materials interact with incident EM radiation as typical conductor, reflecting most part of electromagnetic wave.

Sharp increase in EM absorbance for PMMA-based composites at MWNT loading lower than 2 wt. % may be attributed to absence of reflectance allowing propagation of incident EM wave through the sample and its interaction with conjugated π-system of incorporated carbon nanotubes.

The value of EM absorbance reaches saturation for MWNT loadings higher than 2 wt. %, corresponding to electrical percolation threshold value. This phenomenon indicates formation of interconnected array (or cluster) of carbon nanotubes in the volume of the polymer, providing conductive paths for electrical current, decreasing electrical resistivity of the composite with subsequent increase in EM reflectance. Further increase of EM reflectance with growing MWNT loading is caused by formation of new conductive paths in linked nanotube network, increasing its electrical conductivity and reflective properties, however not affecting or diminishing absorption of EM radiation.

Polystyrene-based composites with same MWNTs as filler show surprisingly high shielding efficiency, as they are almost insulating at low nanotube loadings. In contrast to PMMA-based materials main part of shielding efficiency of PS-based composites is provided by absorption, growing with increase of MWNT concentration, especially for higher frequencies. At the same time reflection of EM radiation is increasing slowly, showing similar behavior as compared with EM absorbance for MWNT/PMMA composites.

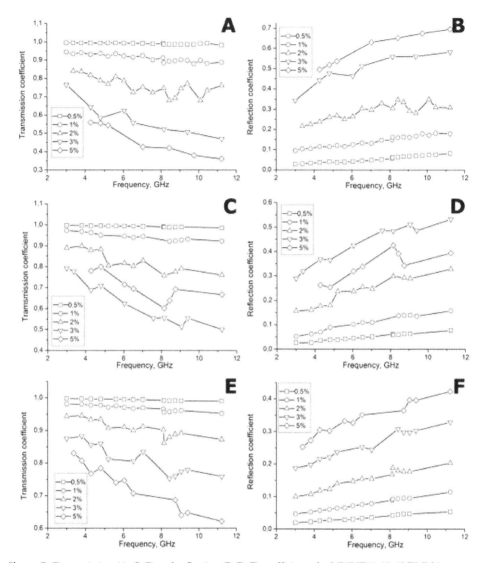

Figure 7. Transmission (A, C, E) and reflection (B, D, F) coefficients for MWNT#1, #2, #3/PMMA composites respectively.

Internal structure of composite materials and interaction between MWNTs and polystyrene matrix show *i)* strong wetting of MWNT surface with polymer, providing certain insulation of nanotubes from each other; *ii)* high dispersion degree of MWNT without primary agglomerates; *iii)* random uniform distribution state of individual nanotubes in the volume of the polymer even for high filler concentrations. All these factors affect changes in EMI shielding mechanism of MWNT/PS composites as compared with relatively poorly-dispersed highly-conductive PMMA-based materials.

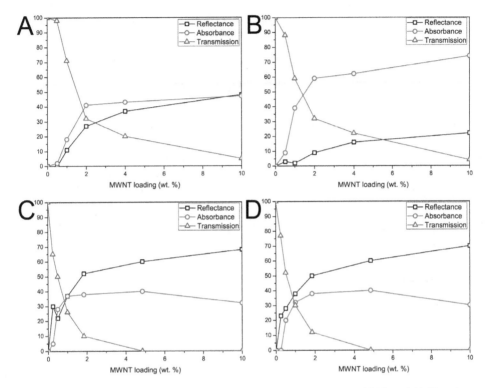

Figure 8. Concentration dependence of electromagnetic response properties of PS-based (A, B) and PMMA-based (C, D) composites in Ka-band. A, C – 29 GHz, B, D – 34 GHz.

Low-conductive PS composites possess correspondingly low EMI reflectance from the surface, allowing electromagnetic radiation to propagate in-depth of the sample. The characteristic wavelength for the EM radiation in microwave region is 1-10 mm, which is higher than the size of the individual nanotube and MWNT agglomerate and is comparable with the macroscopic size of the composite sample. Thus incident EM wave may interact with whole volume of the composite sample and this interaction is strongly facilitated due to the absence of surface conductivity and reflectance due to skin-effects.

Propagating EM wave interacts with individual nanotubes uniformly dispersed in the volume of the composite with corresponding attenuation. High dispersion degree of MWNTs in polymer results in uniform density of attenuating particles, providing correspondingly high "extinction coefficient" of composite material.

3. Conclusion

Properties of multiwall carbon nanotubes and MWNT-containing composite materials are strongly dependent on their structure and morphology and on the type of polymer matrix.

Variations of mean diameter distribution and morphology of agglomerates results in significant changes of their dispersive and electrical properties, affecting electromagnetic response of composites.

MWNT-filled composites with PMMA matrix show higher EM reflectance, higher electrical conductivity with low percolation threshold due to relatively low wetting of hydrophobic MWNTs with polar PMMA molecules and corresponding low dispersion degree of individual nanotubes.

In contrast for low-polar polystyrene matrix high wetting was observed resulting in formation of uniformly dispersed array of PS-covered MWNTs in the matrix. Coverage of MWNT surface with polystyrene results in insulation of nanotubes, thus resulting in high electrical resistivity with high percolation threshold. High shielding efficiency was observed for such composites, provided mainly by absorbance of EM radiation with low-conductive media.

Thus variation of main properties of polymer matrices and incorporated nanotubes allows to obtain composite materials with predictable tailorable properties, which may be used in various applications.

Author details

Ilya Mazov
Boreskov Institute of Catalysis, Novosibirsk, Russia
National Research Technical University "MISIS", Moscow, Russia

Vladimir Kuznetsov
Boreskov Institute of Catalysis, Novosibirsk, Russia

Anatoly Romanenko
Nikolaev Institute of Inorganic Chemistry, Novosibirsk, Russia

Valentin Suslyaev
National Research Tomsk State University, Tomsk, Russia

Acknowledgement

This work was partially supported by RFBR grant #11-03-00351, ISTC project B-1708, and RNP projects #2.1.1/10256, #16.740.11.0016 & #16.740.11.0146.

4. References

[1] J.-P. Salvetat, J.-M. Bonard, N.H. Thomson, A.J. Kulik, L. Forro, W. Benoit, L. Zuppiroli, Mechanical properties of carbon nanotubes, Appl. Phys. A 69, 255–260 (1999)
[2] M. Terrones, Science And Technology Of The Twenty-First Century: Synthesis, Properties, And Applications Of Carbon Nanotubes, Annu. Rev. Mater.Res. (2003) 33 419–501
[3] P.M. Ajayan, O.Z. Zhou, Applications of Carbon Nanotubes in M. S. Dresselhaus, G. Dresselhaus, Ph. Avouris (Eds.): Carbon Nanotubes, Topics Appl. Phys. 80, 391–425 (2001)

[4] D. Srivastava, C. Wei, K. Cho, Nanomechanics of carbon nanotubes and composites, Appl. Mech. Rev. 56 (2003) 215-230

[5] J.R. Stetter, G.J. Maclay Carbon nanotubes and sensors: a review, Advanced Micro and Nanosystems, Vol. 1, edited by: H. Baltes, O. Brand, G.K. Fedder, C. Hierold, J. Korvink, O. Tabata 357-382 (2004)

[6] J. Wang, Carbon-Nanotube Based Electrochemical Biosensors: A Review, Electroanalysis 17 (2005) 7-14

[7] J.W.G. Wilder, L.C. Venema, A.G. Rinzler, R.E. Smalley, C. Dekker Electronic structure of atomically resolved carbon nanotubes Nature 391 (1998) 59-62

[8] Q. Liu, W. Ren, Z.-G. Chen, L. Yin, F. Li, H. Cong, H.-M. Cheng, Semiconducting properties of cup-stacked carbon nanotubes, Carbon 47 (2009) 731-736

[9] R.S. Ruoff, D. Qian, W.K. Liu Mechanical properties of carbon nanotubes: theoretical predictions and experimental measurements C.R. Physique 4 (2003) 993–1008

[10] T. W. Ebbesen, P. M. Ajayan, Large scale synthesis of carbon nanotubes, Nature, 358, 220 (1992).

[11] A. Thess, R. Lee, P. Nikolaev, H. Dai, P. Petit, J. Robert, C. Xu, Y.H. Lee, S.G. Kim, A.G. Rinzler, D.T. Colbert, G.E. Scuseria, D.Tománek, J.E. Fischer, R.E. Smalley Crystalline Ropes of Metallic Carbon Nanotubes, Science 273 (1996) 483-487

[12] B. Kitiyanan, W.E. Alvarez, J.H. Harwell, D.E. Resasco, Controlled production of single-wall carbon nanotubes by catalytic decomposition of CO on bimetallic Co–Mo catalysts, Chemical Physics Letters 317 (2000) 497–503

[13] Y. Murakami, S. Chiashi, Y. Miyauchi, M. Hu, M. Ogura, T. Okubo, S. Maruyama, Growth of vertically aligned single-walled carbon nanotube films on quartz substrates and their optical anisotropy, Chemical Physics Letters 385 (2004) 298–303

[14] G. Zhang, D. Mann, L. Zhang, A. Javey, Y. Li, E. Yenilmez, Q. Wang, J.P. McVittie Y. Nishi, J. Gibbons, H. Dai Ultra-high-yield growth of vertical single-walled carbon nanotubes: Hidden roles of hydrogen and oxygen PNAS 102 (2005) 16141–16145

[15] M. Kumar, Y. Ando, Chemical Vapor Deposition of Carbon Nanotubes: A Review on Growth Mechanism and Mass Production Journal of Nanoscience and Nanotechnology 10 (2010) 3739–3758

[16] Y. Li, X.B. Zhang, X.Y. Tao, J.M. Xu, W.Z. Huang, J.H. Luo, Z.Q. Luo, T. Li, F. Liu, Y. Bao, H.J. Geise, Mass production of high-quality multi-walled carbon nanotube bundles on a Ni/Mo/MgO catalyst, Carbon 43 (2005) 295–301

[17] K.B. Hong, A.A.B. Ismail, M.E. Mahayuddin, A.R. Mohamed, S.H.S. Zeiri, Production of High Purity Multi-Walled Carbon Nanotubes from Catalytic Decomposition of Methane Journal of Natural Gas Chemistry 15 (2006) 266-270

[18] D. Venegoni, Ph. Serp, R. Feurer , Y. Kihn, C. Vahlas , Ph. Kalck, Parametric study for the growth of carbon nanotubes by catalytic chemical vapor deposition in a fluidized bed reactor, Carbon 40 (2002) 1799–1807

[19] M. Corrias, B. Caussat, A. Ayral, J. Durand, Y. Kihn, Ph. Kalck, Ph. Serp, Carbon nanotubes produced by fuidized bed catalytic CVD: first approach of the process, Chemical Engineering Science 58 (2003) 4475 – 4482

[20] S.W. Jeong, S.Y. Son, D.H. Lee, Synthesis of multi-walled carbon nanotubes using Co–Fe–Mo/Al$_2$O$_3$ catalytic powders in a fluidized bed reactor, Advanced Powder Technology 21 (2010) 93–99

[21] J. Wang, H. Kou, X. Liu, Y. Pan, J. Guo, Reinforcement of mullite matrix with multi-walled carbon nanotubes, Ceramics International 33 (2007) 719–722

[22] Y.-T. Jang, S.-I. Moon, J.-H. Ahn, Y.-H. Lee, B.-K. Ju A simple approach in fabricating chemical sensor using laterally grown multi-walled carbon nanotubes, Sensors and Actuators B: Chemical 99 (2004) 118–122

[23] N. Sinha, J. Ma, J.T.W. Yeow, Carbon Nanotube-Based Sensors, Journal of Nanoscience and Nanotechnology 6 (2006) 573–590

[24] N. Maksimova, G. Mestl, R. Schlogl Catalytic activity of carbon nanotubes and other carbon materials for oxidative dehydrogenation of ethylbenzene to styrene, Studies in Surface Science and Catalysis, G.F. Froment and K.C. Waugh (Editors) 133 (2001) 383–389

[25] M.S. Saha, R. Li, X. Sun, High loading and monodispersed Pt nanoparticles on multiwalled carbon nanotubes for high performance proton exchange membrane fuel cells, Journal of Power Sources 177 (2008) 314–322

[26] M.H. Al-Saleh, U. Sundararaj, Electromagnetic interference shielding mechanisms of CNT/polymer composites, Carbon 47 (2009) 1738-1746

[27] T.A. Hilder, J.M. Hill Carbon nanotubes as drug delivery nanocapsules, Current Applied Physics 8 (2008) 258–261

[28] C. Tripisciano, K. Kraemer, A. Taylor, E. Borowiak-Palen Single-wall carbon nanotubes based anticancer drug delivery system, Chemical Physics Letters 478 (2009) 200–205

[29] Makar, J. & Beaudoin, J. Carbon nanotubes and their application in the construction industry. Nanotechnology 331-341 (2003

[30] M. Sangermano, S. Pegel, P. Potschke, B. Voit Antistatic Epoxy Coatings With Carbon Nanotubes Obtained by Cationic Photopolymerization, Macromol. Rapid Commun. 29 (2008) 396–400

[31] Y. Yang, M.C. Gupta, K.L. Dudley, R.W. Lawrence A Comparative Study of EMI Shielding Properties of Carbon Nanofiber and Multi-Walled Carbon Nanotube Filled Polymer Composites Journal of Nanoscience and Nanotechnology 5 (2005) 927-931

[32] X. Zhang, J. Zhang, Z. Liu, Conducting polymer/carbon nanotube composite films made by in situ electropolymerization using an ionic surfactant as the supporting electrolyte, Carbon 43 (2005) 2186–2191

[33] Sun, L. and Sue, H.-J. (2010) Epoxy/Carbon Nanotube Nanocomposites, in Epoxy Polymers: New Materials and Innovations (eds J.-P. Pascault and R. J. J. Williams), Wiley-VCH Verlag GmbH & Co. KGaA, Weinheim, Germany

[34] S. Shang, W. Zeng, X. Tao, High stretchable MWNTs/polyurethane conductive nanocomposites J. Mater. Chem., 21 (2011) 7274-7280

[35] Y. Zou, Y. Feng, L. Wang, X. Liu, Processing and properties of MWNT/HDPE composites, Carbon 42 (2004) 271–277

[36] W. Tang, M.H. Santare, S.G. Advani, Melt processing and mechanical property characterization of multi-walled carbon nanotube /high density polyethylene (MWNT/HDPE) composite films, Carbon 41 (2003) 2779–2785

[37] S. Shang, L. Li, X. Yang, Y. Wei Polymethylmethacrylate-carbon nanotubes composites prepared by microemulsion polymerization for gas sensor, Composites Science and Technology 69 (2009) 1156–1159

[38] J.H. Lehman, M. Terrones, E. Mansfield, K.E. Hurst, V. Meunier, Evaluating the characteristics of multiwall carbon nanotubes, Carbon 49 (2011) 2581–2602

[39] J. Hilding, E.A. Grulke, Z.G. Zhang, F. Lockwood, Dispersion of Carbon Nanotubes in Liquids, Journal Of Dispersion Science And Technology 24 (2003) 1-41

[40] L.S. Schadler, S.C. Giannaris,P.M. Ajayan, Load transfer in carbon nanotube epoxy composites, Appl. Phys. Lett. 73, 3842 (1998)

[41] L.R. Xu, S. Sengupta, Interfacial stress transfer and property mismatch in discontinuous nanofiber/nanotube composite materials, J. Nanosci Nanotechnol. 4 (2005) 620-626

[42] K.T. Kim, J. Eckert, S.B. Menzel, T. Gemming, S.H. Hong, Grain refinement assisted strengthening of carbon nanotube reinforced copper matrix nanocomposites Appl. Phys. Lett. 92, 121901 (2008)

[43] X. Zeng, G. Zhou, Q. Xu, Y. Xiong, C. Luo, J. Wu, A new technique for dispersion of carbon nanotube in a metal melt, Materials Science and Engineering: A 527 (2010) 5335–5340

[44] K. Q. Xiao, L.C. Zhang, Effective separation and alignment of long entangled carbon nanotubes in epoxy, Journal of Materials Science 40 (2005) 6513–6516

[45] S. Bal, S.S. Samal Carbon nanotube reinforced polymer composites–A state of the art, Bull.Mater. Sci. 30 (2007) 379–386

[46] Y.J. Kim, T.S. Shin, H.D. Choi, J.H. Kwon, Y.-C. Chung, H.G. Yoon Electrical conductivity of chemically modified multiwalled carbon nanotube/epoxy composites, Carbon 43 (2005) 23–30

[47] R.E. Gorga, K.K.S. Lau, K.K. Gleason, R.E. Cohen, The Importance of Interfacial Design at the Carbon Nanotube/Polymer Composite Interface, Journal of Applied Polymer Science, 102 (2006) 1413–1418

[48] Z. Yang, Z. Cao, H. Sun, Y. Li Composite Films Based on Aligned Carbon Nanotube Arrays and a Poly(N-Isopropyl Acrylamide) Hydrogel, Adv. Mater. 2008, 20, 2201–2205

[49] E.J. Garcia, A.J. Hart, B.L. Wardle, A.H. Slocum, Fabrication and Nanocompression Testing of Aligned Carbon-Nanotube–Polymer Nanocomposites, Adv. Mater. 2007, 19, 2151–2156

[50] W. Yuan, J. Che, M.B. Chan-Park A Novel Polyimide Dispersing Matrix for Highly Electrically Conductive Solution-Cast Carbon Nanotube-Based Composite, Chem. Mater. 23 (2011) 4149–4157

[51] P. Ciselli, R. Zhang, Z. Wang, C.T. Reynolds, M. Baxendale, T. Peijs, Oriented UHMW-PE/CNT composite tapes by a solution casting-drawing process using mixed-solvents European Polymer Journal 45 (2009) 2741–2748

[52] N.G. Sahoo, Y.C. Jung, H.J. Yoo, J.W. Cho, Effect of Functionalized Carbon Nanotubes on Molecular Interaction and Properties of Polyurethane Composites, Macromol. Chem. Phys. 207 (2006) 1773–1780

[53] K. Zhang, J.Y.Lim , B.J.Park, H.J. Jin, H.J.Choi Carboxylic Acid Functionalized Multi-Walled Carbon Nanotube-Adsorption onto Poly(methyl methacrylate) Microspheres Journal of Nanoscience and Nanotechnology 9 (2009) 1058-1061

[54] L. Cui, N.H. Tarte, S.I. Woo, Synthesis and Characterization of PMMA/MWNT Nanocomposites Prepared by in Situ Polymerization with Ni(acac)2 Catalyst, Macromolecules 2009, 42, 8649–8654

[55] H.-J. Jin, H.J. Choi, S.H. Yoon, S.J. Myung, S.E. Shim, Carbon Nanotube-Adsorbed Polystyrene and Poly(methyl methacrylate) Microspheres Chem. Mater. 2005, 17, 4034-4037

[56] D. Wang, J. Zhu, Q. Yao, C.A. Wilkie, A Comparison of Various Methods for the Preparation of Polystyrene and Poly(methyl methacrylate) Clay Nanocomposites, Chem. Mater.2002, 14, 3837-3843

[57] T. Wang, S. Shi, F. Yang, L. M. Zhou, S. Kuroda, Poly(methyl methacrylate)/polystyrene composite latex particles with a novel core/shell morphology, J Mater Sci (2010) 45:3392–3395

[58] F. Du, J.E. Fischer, K.I. Winey, Coagulation Method for Preparing Single-Walled Carbon Nanotube/Poly(methyl methacrylate) Composites and Their Modulus, Electrical Conductivity, and Thermal Stability, Journal of Polymer Science: Part B: Polymer Physics 41 (2003) 3333–3338

[59] T. Gabor, D. Aranyi, K. Papp, F.H. Karman, E. Kalman, Dispersibility of Carbon Nanotubes, Materials Science Forum 161 (2007) 537-538

[60] A. Usoltseva, V. Kuznetsov, N. Rudina, E. Moroz, M. Haluska, S. Roth, Influence of catalysts' ctivation on their activity and selectivity in carbon nanotubes synthesis, Phys. Stat. Sol. 11 (2007) 3920–3924.

[61] M. Sikka, N.N. Pellegrini, E.A. Schmitt, K.I. Winey, Modifying a Polystyrene/Poly (methyl methacrylate) Interface with Poly(styrene-co-methyl methacrylate) Random Copolymers, Macromolecules 30 (1997) 445 -455

[62] S. Varennes, H. P. Schreiber, On Origins of Time-Dependence in Contact Angle Measurements, The Journal of Adhesion, 76:293-306, 2001

[63] Y. Li, Y. Yang , F. Yu, L. Dong, Surface and Interface Morphology of Polystyrene/Poly (methyl methacrylate) Thin-Film Blends and Bilayers, Journal of Polymer Science: Part B: Polymer Physics, Vol. 44, 9–21 (2006)

[64] Y.T. Sung, W.J. Seo, Y.H. Kim, H.S. Lee, W.N. Kim Evaluation of interfacial tension for poly(methyl methacrylate) and polystyrene by rheological measurements and interaction parameter of the two polymers, Korea-Australia Rheology Journal 16 (2004) 135-140

[65] Tanaka K, Takahara A, Kajiyama T. Macromolecules 1996;29:3232-9.

[66] C. Ton-That, A.G. Shard, D.O.H. Teare, R.H. Bradley, XPS and AFM surface studies of solvent-cast PS/PMMA blends, Polymer 42 (2001) 1121-1129

[67] S. Wu, Polar and Nonpolar Interactions in Adhesion, J. Adhesion, 1973. Vol. 5, pp. 39-55

[68] P.K.C. Pillai, P. Khurana, A. Tripathi, Dielectric studies of poly(methyl methacrylate) /polystyrene double layer system Journal of Materials Science Letters 5 (1986) 629 – 632

[69] I. Mazov, V.L. Kuznetsov, I.A. Simonova, A.I. Stadnichenko, A.V. Ishchenko, A.I. Romanenko, E.N. Tkachev, O.B. Anikeeva "Oxidation behavior of multiwall carbon nanotubes with different diameters and morphology", Applied Surface Science, 258, 17 (2012) 6272–6280

[70] L. Ci, J. Suhr, V. Pushparaj, X. Zhang, P. M. Ajayan, Continuous Carbon Nanotube Reinforced Composites, Nano Letters 8 (2008) 2762-2766

[71] A. Moisala, Q. Li, I.A. Kinloch, A.H. Windle, Thermal and electrical conductivity of single- and multi-walled carbon nanotube-epoxy composites, Compos. Sci. Technol., 66, 10 (2006), pp 1285-1288

[72] W.K. Park, J.H. Kim Effect of Carbon Nanotube Pre-treatment on Dispersion and Electrical Properties of Melt Mixed Multi-Walled Carbon Nanotubes / Poly(methyl methacrylate) Composites, Macromolecular Research, Vol. 13, No. 3, pp 206-211 (2005)

Chapter 4

C/Li$_2$MnSiO$_4$ Nanocomposite Cathode Material for Li-Ion Batteries

Marcin Molenda, Michał Świętosławski and Roman Dziembaj

Additional information is available at the end of the chapter

1. Introduction

Technological development of portable devices, e.g, mobile phones, laptops, etc., as well as progress in electrical vehicles (EV) and hybrid electrical vehicles (HEV) technologies require batteries efficient in volumetric and gravimetric energy storage, exhibiting large number of charge/discharge cycles and being cheap and safe for users. Moreover, materials used in energy storage and conversion systems should be environmentally friendly and recyclable. Currently, rechargeable lithium-ion batteries (LIBs) are the most popular portable energy storage system, mostly due to their highest energy density among all others rechargeable battery technologies, like Ni-Cd or Ni-MH cells which reached the theoretical limit of performance. Commercially available LIBs are based on layered lithium cobalt oxide (LiCoO$_2$) or related systems, which are expensive and toxic. These materials are unstable in an overcharged state, thus the battery safety is affected, especially in high power (20-100 kWh) applications for EV, HEV and renewable energy systems. The bigger battery capacity results in more energy accumulated, thus operational safety is a key issue. On the other hand, lifetime and capacity retention of LIBs in changeable operation conditions (from -30°C to +60°C, average lifetime 2-4 years) are a challenges to develop new materials and cell assembly technologies.

2. Li-ion battery technology

First rechargeable lithium cells taking advantages of intercalation process were developed in year 1972 [1]. The cells Li/Li$^+$/Li$_x$TiS$_2$ revealed 2V potential and relatively low gravimetric capacity. Applications of metallic lithium as anode material resulted in common cell breakdown due to formation of dendritic structures on anode during cell cycling. The problems forced research and development of new intercalation materials for lithium batteries. In the eighties a new conception of lithium cell was proposed (so called Li-ion batteries or "rocking-chair batteries"), which consisted of application of two different lithium intercalation compounds as anode and cathode materials [2-4]. As anode material a graphite intercalated with lithium was used while cathode materials were based on layered 3d transition metal oxides.

2.1. Layered LiCoO$_2$ oxide and related systems

The first Li$_x$C$_6$/Li$^+$/Li$_{1-x}$CoO$_2$ Li-ion battery system was commercialized in 1993 by Sony Co. In Fig. 1 a working mechanism during discharge cycle of Li-ion cell is presented.

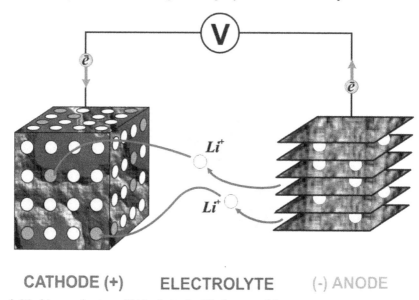

CATHODE (+)   ELECTROLYTE   (-) ANODE

Figure 1. Working mechanism of Li-ion batteries (discharge cycle).

The electrochemical reaction at the graphite anode side can be written as (1):

$$C_6 + xLi^+ + xe^- \leftrightarrow Li_xC_6 \tag{1}$$

and suitably at LiCoO$_2$ cathode side (2):

$$LiCoO_2 \leftrightarrow Li_{1-x}CoO_2 + xLi^+ + xe^- \tag{2}$$

Commercially available LIBs based on layered lithium cobalt oxide (LiCoO$_2$) or related systems (LiNi$_y$Co$_{1-y}$O$_2$, LiMn$_y$Co$_{1-y}$O$_2$, LiMn$_{1/3}$Ni$_{1/3}$Co$_{1/3}$O$_2$) reveal reversible capacity 130-150 mAh/g and working potential 3.6-3.7 V. The oxide materials are expensive and toxic due to cobalt content and are unstable in an overcharged state, thus the safety is strongly affected. This is related to strong oxidizing behavior of the charged layered oxide cathode in contact with organic electrolyte what may lead to combustion or even explosion [4, 5]. Unfortunately, this effect may be also increased by application of nanosized materials with high surface area.

2.2. Spinel LiMn$_2$O$_4$ and related systems

An alternative material for cathode based on layered oxides LiM$_y$Co$_{1-y}$O$_2$ (M=Ni, Mn, Fe) is LiMn$_2$O$_4$ spinel. LiMn$_2$O$_4$ reveals a little lower reversible capacity 120 mAh/g at working potential 4 V, but the material is distinctly cheaper and nontoxic. However, application of spinel as cathode material in commercial Li-ion batteries is retarded by phase transition observed near room temperature, i.e. battery operation temperature, and related to Jahn-Teller distortion of Mn^{3+} ions, what resulted in capacity fading. Stabilization of cubic spinel structure is possible by controlled formation of cationic defects [6], by lithium [7] or 3d metal substitution [8] into spinel lattice as well as by isoelectronic sulfur substitution [9, 10]. Instability of high oxidation state of transition metal in spinel structure observed in charged state of cathode leads to oxygen evolution, similarly to the layered oxide cathodes, and reaction with electrolyte. Application of spinel based materials is limited to cheap battery packs for EV [11].

2.3. Olivine LiFePO$_4$ cathode material

An interesting and very promising group of insertion materials are LiMXO$_4$ (M= metal 3d, X=S, P, As, Mo, W) type compounds [12], with LiFePO$_4$ among them. LiFePO$_4$ lithium iron phosphate of olivine structure is chemically and thermally stable material with relatively high gravimetric capacity 170 mAh/g at working potential 3.5 V. The material is cheap, nontoxic and environmental friendly. Its high chemical stability towards electrolyte, related to strong P-O bonds, significantly improves safety of LIBs. However, very low electrical conductivity of LiFePO$_4$ system (10^{-9} S/cm @RT) requires application of carbon coatings and composite formation [13, 14].

3. Orthosilicates Li$_2$MSiO$_4$ – New high capacity cathode materials

Application of orthosilicates Li$_2$MSiO$_4$ (M=Fe, Mn, Co, Ni) compounds as insertion materials for LIBs was firstly proposed by Prof. Goodenough [12, 15]. The materials mainly crystallizes in orthorhombic system of Pmn21 space group in olivine structure [16-18]. The Li$_2$MSiO$_4$ olivine structure may be described as wavy layers of [SiMO$_4$]$_\infty$ on *ac* axis plane and connected along *b* axis by LiO$_4$ tetrahedra [16]. Within the layers every tetrahedron SiO$_4$ shares its corner with four next MO$_4$ tetrahedra. Lithium ions occupy the tetrahedral sites (LiO$_4$) between two layers and share 3 and 1 oxygen atoms with the layers. In fact, diffusion of lithium ions in this structure is possible only through the canals formed by LiO$_4$ tetrahedra. Li$_2$MSiO$_4$ silicates reveal a possibility of reversible insertion of two lithium ions per molecule, so this leads to exchange of two electrons by transition metal. As a results the silicates reveal very high theoretical gravimetric capacity, up to 330 mA/g. However, structural limitation due to sharing of LiO$_4$ between two wavy layers, should affect a little the capacity. Calculations of electrochemical potential of lithium insertion/deinsertion show a two stage process:

$$Li_2M^{2+}SiO_4 \rightarrow \square LiM^{3+}SiO_4 + Li^+ \qquad (3)$$

$$\Upsilon LiM^{3+}SiO_4 \rightarrow \Upsilon_2 M^{4+}SiO_4 + Li^+ \qquad (4)$$

Depending on transition metal potential of the reaction (3) vary from 3.2 V (for Fe) to 4.1-4.4 V (for Mn, Co, Ni), while the potential of reaction (4) is in range 4.5-5.0 V. Deinsertion of the second lithium ion requires applying high potential above 4.5 V, and this is a challenge for electrolyte. Orthosilicates Li_2MSiO_4 (M=Fe, Mn, Co, Ni) materials reveal very low electrical conductivity (10^{-12} - 10^{-15} S/cm @RT) and carbon coating of the materials is required. On the other hand, downsizing of material grains should improve electrochemical performance [18]. Very strong covalent bonding of Si-O results in high chemical stability towards electrolyte, strongly increasing the safety of LIBs based on silicates. The materials are nontoxic, environmental friendly and cheap. The properties mentioned above provide a challenge for developing a composite cathode materials based on Li_2MSiO_4 orthosilicates.

4. Li$_2$MnSiO$_4$ nanostructured cathode material

Li_2MnSiO_4 is a member of dilithiumorthosilicates Li_2MSiO_4 (M = Fe, Mn, Co) family, thanks to strong covalent Si-O bond, it shows high thermal and chemical stability. High theoretical capacity 333 mAh/g, low production costs, safety and optimal working potential make them an attractive cathode material. Very low electrical conductivity (10^{-12} - 10^{-15} S/cm @RT) can be improved by coating with conductive carbon layers (CCL) and by grains size downsizing [18-20]. The properties of Li_2MnSiO_4 material comes to a conclusion that cathode material for LIBs based on this compound should be prepared as C/Li_2MnSiO_4 nanocomposite. Thus, the special preparation techniques have to be applied in terms to obtain nanocrystaline grains of Li_2MnSiO_4 cathode material coated by conductive carbon layers (CCL). During last few years several different technics of Li_2MnSiO_4 synthesis were proposed. Hydrothermal synthesis [21-24] as well as solid-state reactions [25-28] can lead to one phase product but the control of the grain size is significantly limited. Sol-gel method is one of the soft chemistry techniques which can be used in synthesis of nanosized lithium orthosilicates [29-33]. Sol-gel processes, especially Pechini's method is a simple technic, characterized by low cost and low temperature of treatment, resulting in homogenous, high purity materials.

4.1. Preparation of nanostructured Li$_2$MnSiO$_4$

Li_2MnSiO_4 was produced using sol-gel synthesis – Pechini type. Starting reagents were: lithium acetate dihydrate (Aldrich), manganese acetate tetrahydrate (Aldrich), tetraethoxysilane (TEOS) (98%, Aldrich) as a source of silicon, ethylene glycol (POCh), citric acid (POCh) and ethanol (POCh). Thanks to chelating metal ions in solution by citric acid the cations can be mixed at the molecular level and the stoichiometric composition can be achived. The reactants were mixed in a molar ratio 1:1:18:6:4:16 - $Mn:Si:C_2H_6O_2:C_6H_8O_7:C_2H_5OH:H_2O$. Based on previous studies it was affirmed, that using 20% excess of lithium acetate leads to one-phase product. All reagents were dissolved in glass reactor under constant argon flow (Ar 5.7). As a solvent, only stoichiometric amount of distilled water was used. Heating water to 35 °C assure fast and complete dissolution of metal acetates. Prepared mixture was heated to 60 °C and few drops of concentrated hydrochloric acid was added to initiate polymerization of metal citrates using ethylene glycol and TEOS. Reaction was conducted for 24h in close reactor. Obtained gel was aged for 3 days at 60 °C in close reactor (Ar atmosphere) and for 3 days at 60 °C in an air-drier (Air atmosphere).

Thermogravimetric analysis (TGA) coupled with simultaneous differential thermal analysis (SDTA) and mass spectrometry evolved gas analysis (MS-EGA) of precursor were performed in Mettler-Toledo 851ᵉ thermo-analyzer using 150 µl corundum crucibles under flow of air/argon (80 ml/min), within temperature range 20–800 °C with heating rate of 10 °C/min. The simultaneous MS-EGA was performed in on-line joined quadruple mass spectrometer (QMS) (Thermostar-Balzers). The 17, 18 and 44 m/z mass lines, ascribed to OH, H_2O and CO_2 species respectively, were collected during the TGA experiments. TGA of the Li_2MnSiO_4 precursor are shown in the Fig. 2 and Fig. 3.

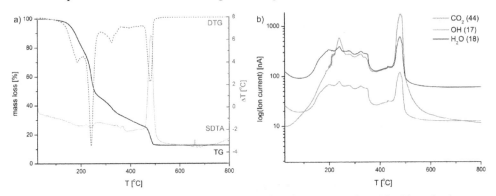

Figure 2. Thermal analysis (TGA/DTG/SDTA) conducted under air atmosphere (a) with evolved gas analysis (EGA-QMS) (b) of Li_2MnSiO_4 precursor.

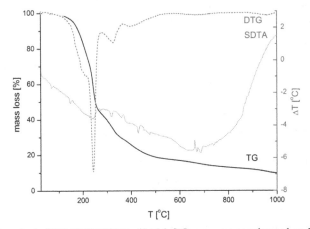

Figure 3. Thermal analysis (TGA/DTG/SDTA) of Li_2MnSiO_4 precursor performed under inert atmosphere (Ar 5.0).

According to TGA curve (Fig. 2a) a complete decomposition of gel organic matrix occurs above 500 °C. Evolved gas analysis confirmed the disintegration of organic components. Amount of active material can be estimated at about 12 wt.% of the precursor. Basing on TGA results calcination conditions were chosen. Li$_2$MnSiO$_4$ precursor was calcined under Ar flow at 600, 700, 800 and 900 °C.

4.2. Properties of nanostructured Li$_2$MnSiO$_4$

X-ray powder diffraction patterns of the samples were collected on BRUKER D2 PHASER using Cu K$_\alpha$ radiation = 1.5418 Å. In Fig. 4 XRD patterns of Li$_2$MnSiO$_4$ obtained at different temperatures (600, 700, 800 and 900 °C) are collected.

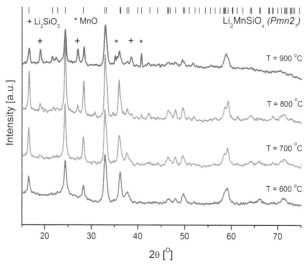

Figure 4. XRD patterns of Li$_2$MnSiO$_4$ after calcination at 600 °C, 700 °C, 800 °C and 900 °C. Li$_2$MnSiO$_4$ (*Pmn2$_1$* phase) diffraction lines positions are marked in the top of the figure.

In the sample calcined at 600 °C all diffraction lines can be attributed to pure Li$_2$MnSiO$_4$ phase (*Pmn2$_1$*). In other specimens obtained at higher temperatures, MnO and Li$_2$SiO$_3$ impurities starts to appear. Formation of MnO and Li$_2$SiO$_3$ in samples calcined at temperatures above 700 °C can be connected, with Li$_2$MnSiO$_4$ decomposition. Average crystallites sizes calculated from diffraction lines broadening based on Scherrer equation are in the range of 17-19 nm depending on calcination temperature of the sample (Tab. 1). The calcination under carefully selected conditions leads to a transformation of a part of organic compounds from the gel matrix into thick carbon layer covering the active material. The temperature programmed oxidation (TPO) performed using TGA equipment confirmed presence of carbon in samples. Fig. 5b presents exemplary TG/DTG/SDTA curves from TPO conducted under air flow in temperature range of 20-1000 °C. The amounts of carbon in samples calcined in different temperatures were calculated from TPO measurements and are collected in Table 1.

Measurements of the specific surface area of the samples were performed in Micrometrics ASAP 2010 using BET isotherm method. About 500 mg of each sample was preliminary degassed at 250–300 °C for 3 h under pressure 0.26–0.4 Pa. Then, N_2 sorption was performed at pressure of $8 \cdot 10^4$ Pa. One of pore size distribution plot with the adsorption-desorption curves inset is presented in Fig. 5a. Exact values of specific surface area and average pores diameters are presented in table 1.

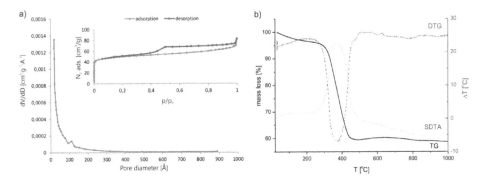

Figure 5. Adsorption-desorption isotherm and pore diameter distribution of C/Li₂MnSiO₄ calcined at 600°C, b) TGA/DTG/SDTA curves of C/Li₂MnSiO₄ calcined at 600 °C.

Low temperature N_2-adsorption (BET) measurements of C/Li₂MnSiO₄ composites show that specific surface area of samples decrease with increasing calcination temperature. Decrease of specific surface area is connected with graphitization of carbon during calcination at higher temperatures.

Sample name	Precursor calcination temperature	Carbon content	Specific surface area (BET)	Average pore dimension (BJH)	Crystallite size (from XRD)
C/Li₂MnSiO₄@600	600 °C	34%	169 m²/g	50 Å	17 nm
C/Li₂MnSiO₄@700	700 °C	32%	152 m²/g	72 Å	18 nm
C/Li₂MnSiO₄@800	800 °C	32%	144 m²/g	73 Å	19 nm
C/Li₂MnSiO₄@900	900 °C	35%	57 m²/g	73 Å	19 nm

Table 1. Morphology parameters of C/Li₂MnSiO₄ composites calcined at different temperatures.

Electrical conductivity was measured using the AC (33Hz) 4-probe method within temperature range of -40÷55 °C. The carbon coated composite powders were so elastic that the standard preparation of pellets was impossible. The fine powder samples were placed into a glass tube and pressed by a screw-press between parallel gold disc electrodes (∅=5 mm) till the measured resistance remained constant. The results of electrical conductivity measurements are gather in Fig. 6. All composites exhibit good electrical conductivity (up to 0,36 S/cm for C/Li₂MnSiO₄@900) in comparison with Li₂MnSiO₄ itself ($\sim 10^{-15}$ S/cm at room temperature - RT). It can be observed that temperature dependence of conductivity for composites calcined at higher temperatures (700, 800 and 900 °C) seem unaffected, by temperature (metallic-like behavior). Conductivity value and its invariability against temperature indicate that carbon layers consist of graphite-like domains. In case of C/Li₂MnSiO₄@600°C the carbon layers is more disordered and probably consist of an activated-like carbon. Those results are consistent with BET measurements in case of graphitization process occurring at higher temperatures.

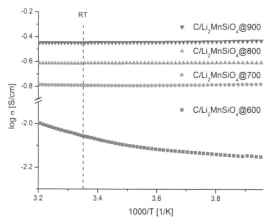

Figure 6. Electrical conductivity measurement of C/Li₂MnSiO₄ (temp. range: -20 to 40°C). Dash line represent room temperature (RT).

Selected samples were investigated using transmission electron microscope (TEM) (Fig. 7 and Fig. 8). Micrographs were collected on TECNAI G2 F20 (200 kV) coupled with an energy dispersive X-ray spectrometer (EDAX).

TEM micrographs reveal well dispersed lithium manganese orthosilicate grains (dark grey and black dots) in the carbon matrix. Figures 7a; 7c; 7d present C/Li₂MnSiO₄@600 at different magnifications. Fig. 7b shows STEM-HAADF (*Scanning Transmission Electron Microscopy – High Angular Annular Dark Field*) image of the C/Li₂MnSiO₄@600 composite. Energy dispersive X-ray spectroscopy (EDS) (Fig. 7b1; 7b2) confirmed presence of lithium manganese silicate in the spots marked in the micrograph 7b (Cu peaks in all EDS spectra originate from TEM copper grid). High resolution electron microscopy image (HREM) shows single grain of Li₂MnSiO₄ (Fig. 7e). Corresponding to that IFFT (*Inverse Fast Furrier Transform*) image (Fig. 7f) clearly shows a crystalline structure of obtained material which can be identified as Li₂MnSiO₄. Grains in the composite have sizes in a range of 5-10 nm. Microstructure of a C/Li₂MnSiO₄ composite calcined at 700 °C is shown in Fig. 8. C/Li₂MnSiO₄@700 °C composite has a bimodal distribution of a crystallite size. TEM micrographs show presence of small (5-10 nm) Li₂MnSiO₄ crystallites, analogues of C/Li₂MnSiO₄@600 °C and a bigger, well crystallized silicates grains, in a range of 50-150 nm (Fig. 8a, c, d). Both types of crystallites in the sample have the same composition which is confirmed by EDS analysis presented in fig. 8b1, b2, b3.

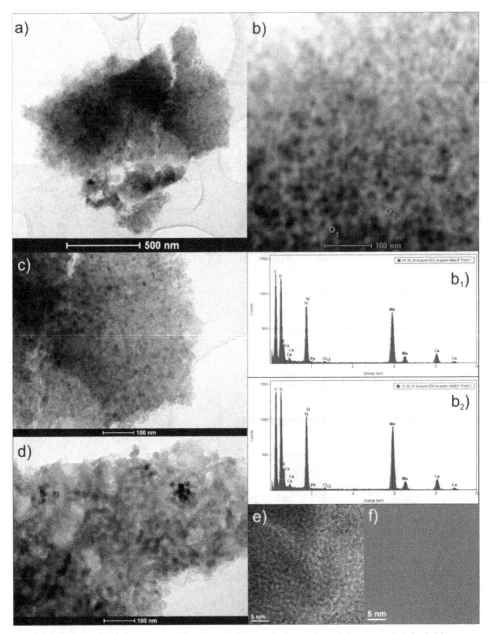

Figure 7. TEM micrographs of C/Li₂MnSiO₄@600. Micrographs a), c) and d) present bright field images of C/Li₂MnSiO₄@600 composite in different magnification; b) microstructure image observed in STEM-HAADF with EDS analysis (b₁; b₂) from marked points; e) high resolution micrograph (HREM) of a single grain and IFFT image f) from the same region.

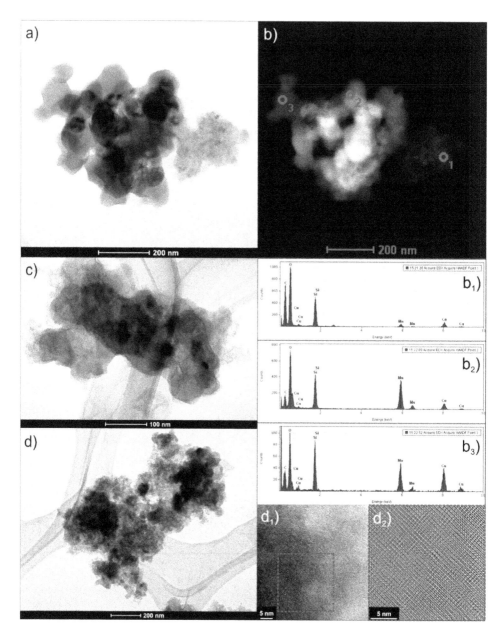

Figure 8. TEM micrographs of C/Li₂MnSiO₄@700. Micrographs a), c) and d) presents bright field images of C/Li₂MnSiO₄@700 composite in different magnification; b) microstructure image observed in STEM-HAADF with EDS analysis (b₁; b₂; b₃) from selected points; d₁) high resolution micrograph (HREM) of region marked in Fig. 8d; d₂) IFFT image from square marked in d₁).

STEM-HAADF image (Fig. 8b) from the same region as a bright field Fig. 8a shows points from where EDS analysis was conducted. All of lithium manganese orthosilicate crystallites in C/Li$_2$MnSiO$_4$@700 °C are well dispersed and fully covered in amorphous carbon. HREM image in Fig. 8d$_1$ with IFFT in Fig. 8d$_2$ shows crystalline structure of Li$_2$MnSiO$_4$ grain.

5. C/Li$_2$MnSiO$_4$ composite cathode material

Nevertheless obtained composites show homogeneous distribution of particles in carbon matrix and exhibit good electrical conductivity they work poorly in a battery cell (see paragraph 5.2). Thickness of primary carbon coating on active material grains strongly limits the electrochemical performance of composite.

5.1. CCL/Li$_2$MnSiO$_4$ composite preparation and properties

Preparation of composites with well define morphology of carbon layers and optimal carbon content was achieved by burning out the primary carbon and recoating Li$_2$MnSiO$_4$ nanosized grains with conductive carbon layers (CCL/Li$_2$MnSiO$_4$ composites). Burning out of primary carbon formed in the sample during synthesis process was carried through calcination of C/Li$_2$MnSiO$_4$@600 and C/Li$_2$MnSiO$_4$@700 under air flow at 300 °C for 3 h. CCL/Li$_2$MnSiO$_4$ composites were produced by wet polymer precursor deposition on active material grains and subsequent controlled pyrolysis [14, 20, 34]. Poly-N-vinylformamide (PNVF) obtained by radical-free polymerization of N-vinylformamide (Aldrich) with pyromellitic acid (PMA) additive (5-10 wt%) was used as a carbon polymer precursor [34]. To achieve an impregnation, Li$_2$MnSiO$_4$ grains were suspended in water polymer solution (8-15 wt%). Finally, samples were dried up in an air drier at 90 °C for 24 h. Prepared samples were pyrolyzed at 600 °C for 6 h under inert atmosphere (Ar).

Burning out primary carbon from the surface of lithium manganese silicate in air atmosphere leads to partial decomposition of active material. Fig. 9 shows XRD patterns of C/Li$_2$MnSiO$_4$@600, Li$_2$MnSiO$_4$@600 without carbon and CCL/Li$_2$MnSiO$_4$@600 composite after coating with CCL from polymer precursor.

In diffraction patterns of Li$_2$MnSiO$_4$@600 it is clearly visible that after burning out primary carbon layer new phase appears in the sample (LiMn$_2$O$_4$). Proposed carbon coating process can reverse decomposition of Li$_2$MnSiO$_4$. After coating Li$_2$MnSiO$_4$ (*Pmn2$_1$*) is a dominant phase in CCL/Li$_2$MnSiO$_4$@600 and all the lines, except for very low intensity line around 19°, can be associated with Li$_2$MnSiO$_4$ phase. CCL preparation procedure influence active material grain size. Average crystallite size calculated using Scherer equation in CCL/Li$_2$MnSiO$_4$@600 is ca. 25 nm. TEM analysis confirms increase of crystallites diameters. Li$_2$MnSiO$_4$ grains observed under TEM are in the range 25-75 nm. TEM micrographs of CCL/Li$_2$MnSiO$_4$@600 are presented in Fig. 10.

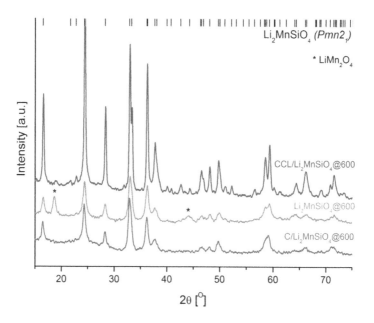

Figure 9. XRD patterns of C/Li₂MnSiO₄@600, Li₂MnSiO₄@600 and CCL/Li₂MnSiO₄@600.

Bright field micrographs (Fig. 10a, c, d) shows crystallites of Li₂MnSiO₄ covered with conductive carbon layers (CCL). Formed carbon coatings adhere well to the surface of active material grains, no voids are visible at the CCL/Li₂MnSiO₄ interface (Fig. 10e). Silicate grains are uniformly covered with approximately 4-5 nm thick CCL. EDS analysis once again confirm presence of lithium manganese silicate in the composite (Fig. 10b, b₁). Carbon content in samples was calculated from TPO measurements and they are displayed in Table 2.

Sample name	Precursor calcination temperature	Carbon content
CCL/Li₂MnSiO₄@600_7.5%	600 °C	7.5%
CCL/Li₂MnSiO₄@600_12%	600 °C	12%
CCL/Li₂MnSiO₄@700_5%	700 °C	5%
CCL/Li₂MnSiO₄@700_10%	700 °C	10%

Table 2. Carbon content of CCL/Li₂MnSiO₄ composites.

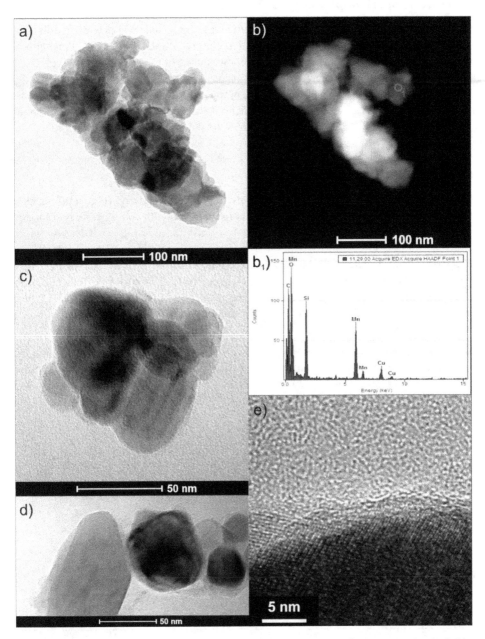

Figure 10. TEM micrographs of CCL/Li$_2$MnSiO$_4$@600. Micrographs a), c) and d) presents bright field images of CCL/Li$_2$MnSiO$_4$@600 composite; b) microstructure image observed in STEM-HAADF with EDS analysis (b$_1$) from selected point; e) HREM micrograph of carbon coating on the active material grain surface.

5.2. Electrochemical properties of C/Li2MnSiO4 composite cathode

The charge-discharge cycling studies of Li/Li+/(C/Li2MnSiO4) cells were conducted in a four electrode configuration using CR2032 assembly between 2.7 and 4.7 V at C/200 rate at room temperature. LiPF6 solution 1M in EC/DEC (1:1) was used in the cells as an electrolyte. The galvanostatic measurements were carried out using ATLAS 0961 MBI test system. Charge-discharge tests were conducted for all CCL composites listed in table 2, composites with primary carbon (C/Li2MnSiO4@600, C/Li2MnSiO4@700 and C/Li2MnSiO4@800) and standard composites (Li2MnSiO4@600_CB and Li2MnSiO4@700_CB). Standard composites were prepared by mixing Li2MnSiO4 powder with commercial carbon additive – carbon black (CB). 15 wt.% of carbon was used. Results collected from charge/discharge tests are presented in fig. 11-16.

C/Li2MnSiO4@600 sample did not show any reversible capacity (Fig. 11a). Lack of electrochemical activity of C/Li2MnSiO4@600 is connected with too high carbon loading. High carbon content (34%) is responsible for limiting of ionic conductivity in the composite and for surface polarization. Galvanostatic cycling studies of C/Li2MnSiO4@700 and C/Li2MnSiO4@800 revealed reversible capacity in a range of 30 mAh·g-1. Fig. 11b show charge and discharge capacity for 7 cycles of C/Li2MnSiO4@700 composite. After first four cycles reversible capacity stabilizes at about 30 mAh·g-1. Very poor performance of this sample is also connected with high amount of carbon in prepared composite (32%).

Standard composites obtained by mixing active silicate materials with carbon black (15%) show extremely low reversible capacity as well (Fig. 12a and 12b). In this case, carbon additive does not provide sufficient electrical contact between active material grains. Due to this fact, even under low C/200 rate samples performed very poorly in charge/discharge tests.

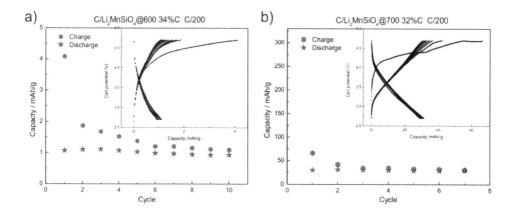

Figure 11. Cell cycling behavior of: a) C/Li2MnSiO4@600, b) C/Li2MnSiO4@700.

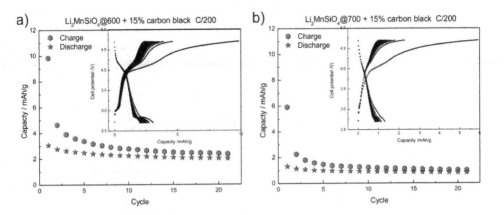

Figure 12. Cell cycling behavior of standard composites Li2MnSiO4 + carbon black (15% carbon content)

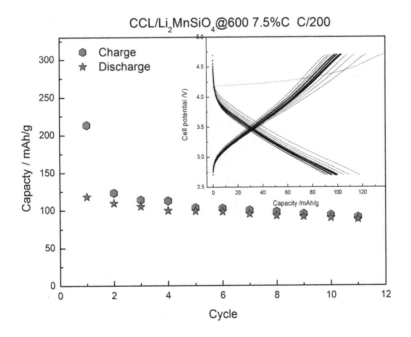

Figure 13. Cell cycling behavior of CCL/Li2MnSiO4@600 (7.5% carbon content)

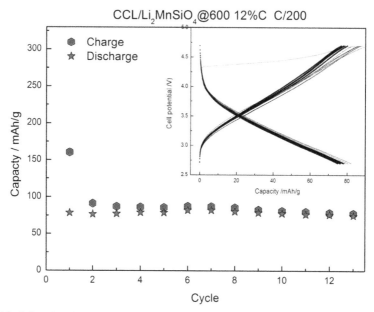

Figure 14. Cell cycling behavior of CCL/Li2MnSiO4@600 (12% carbon content)

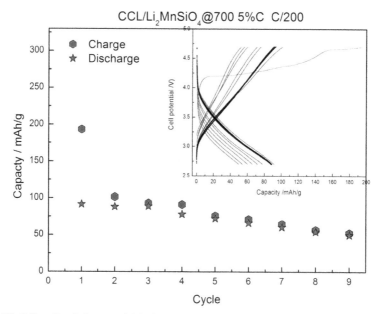

Figure 15. Cell cycling behavior of CCL/Li2MnSiO4@700 (5% carbon content)

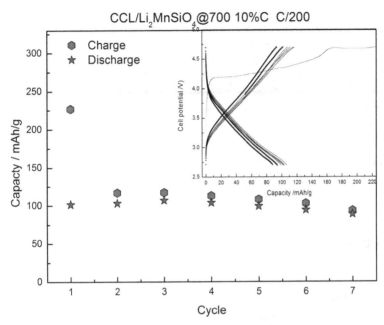

Figure 16. Cell cycling behavior of CCL/Li₂MnSiO₄@700 (10% carbon content)

Fig. 13-16 show cycling behavior of CCL composites with different amounts of carbon. Delithation of CCL composites after initial charging is in the range of 48-68% (charged CCL/Li₂MnSiO₄@600_12%C – correspond to composition Li₁.₀₄MnSiO₄ and charged CCL/Li₂MnSiO₄@700_10%C – correspond to composition Li₀.₆₄MnSiO₄). The observed capacity loss in the first cycles is related to SEI formation in CCL. During lithium extraction, Mn^{3+} and Mn^{4+} appear, and at the same time Li₂MnSiO₄ undergoes decomposition caused by Jahn-Teller distortion associated with changes in lattice parameters during $Mn^{3+} \rightarrow Mn^{4+}$ transition [30]. Despite the fact that crystalline structure collapses, discharge capacity is around 90-110 mAh/g. CCL composites produced from Li₂MnSiO₄ calcined at 600 °C exhibit better coulombic efficiency than Li₂MnSiO₄ calcined at 700 °C, the reversible capacity after 10 cycle is close to 100 mAh/g. The results indicate that coulombic efficiency of cathode material depends on grain size and homogeneity in grain size distribution, while cell capacity is limited by carbon coating performance.

6. Conclusions

Dilithium manganese orthosilicate – a high energy density cathode material – was successfully synthesized by sol-gel Pechini method. Encapsulation of nanosized grains of Li₂MnSiO₄ in carbon matrix, resulted from organic precursor, avoided further sintering. Different crystallite size was obtained, in particular, nanoparticles within range of 5-10 nm. High chemical stability of this material under a highly reductive environment was observed. Application of carbon coating improved electrical conductivity of cathode material to the satisfactory level of ~$10^{-0.5}$ S/cm. Thickness of carbon coating on active material grains strongly limits the electrochemical performance of composite. Formation of CCL/Li$_2$MnSiO$_4$ composites significantly improved electrochemical performance of cathode materials, showing a reversible capacity of 90-110 mAh/g after 10 cycles. Electrochemical tests indicated that composite preparation should be optimized in terms of carbon content, CCL performance and homogeneity of a crystallite size. It was found that coulombic efficiency of cathode material depends on grain size and homogeneity in grain size distribution, while cell capacity is limited by carbon coating performance.

Author details

Marcin Molenda*, Michał Świętosławski and Roman Dziembaj
Jagiellonian University, Faculty of Chemistry, Krakow, Poland

Acknowledgement

This work has been financially supported by the Polish National Science Centre under research grant no. N N209 088638, and by the European Institute of Innovation and Technology under the KIC InnoEnergy NewMat project. The part of the measurements was carried out with the equipment purchased thanks to the financial support of the European Regional Development Fund in the framework of the Polish Innovation Economy Operational Program (POIG.02.01.00-12-023/08). One of the authors (M.Ś.) acknowledge a financial support from the International PhD-studies programme at the Faculty of Chemistry Jagiellonian University within the Foundation for Polish Science MPD. TEM analysis were carried out in Laboratory of Transmission Analytical Electron Microscopy at the Institute of Metallurgy and Material Science, Polish Academy of Sciences.

7. References

[1] Whittingham M.S. Electrical Energy Storage and Intercalation Chemistry. Science 1976;192 1126-1127.

[2] Nagaura T., Tozawa K. Lithium ion rechargeable battery. *Progress in Batteries & Solar Cells* 1990;9 209-217.

[3] Tarascon J.-M., Guyomard D. The Li$_{1+x}$Mn$_2$O$_4$/C rocking-chair system: a review Electrochimica Acta 1993;38(9), 1221-1231.

[4] Tarascon J.-M., Armand M. Issues and challenges facing rechargeable lithium batteries. Nature 2001;414(6861) 359-367.

[5] Zhang Z., Gong Z., Yang Y. Electrochemical Performance and Surface Properties of Bare and TiO$_2$-Coated Cathode Materials in Lithium-Ion Batteries. The Journal of Physical Chemistry B 2004;108(45) 17546–17552.

---

* Corresponding Author

[6]   Molenda M., Dziembaj R., Podstawka E., Proniewicz L.M. Changes in local structure of lithium manganese spinels (Li:Mn=1:2) characterised by XRD, DSC, TGA, IR, and Raman spectroscopy. Journal of Physics and Chemistry of Solids 2005;66(10) 1761-1768.

[7]   Dziembaj R., Molenda M. Stabilization of the spinel structure in $Li_{1+\delta\,7}Mn_{2-\delta\,7}O_4$ obtained by sol–gel method. Journal of Power Sources 2003;119–121C 121-124.

[8]   Molenda J., Marzec J., Świerczek K., Ojczyk W., Ziemnicki M., Wilk P., Molenda M., Drozdek M., Dziembaj R. The effect of 3d substitutions in the manganese sublattice on the charge transport mechanism and electrochemical properties of manganese spinel. Solid State Ionics 2004;171(3-4) 215-227.

[9]   Molenda M., Dziembaj R., Podstawka E., Proniewicz L.M., Piwowarska Z. An attempt to improve electrical conductivity of the pyrolysed carbon-$LiMn_2O_{4-y}S_y$ ($0 \leq y \leq 0.5$) composites. Journal of Power Sources 2007;174(2) 613-618.

[10] Molenda M., Dziembaj R., Podstawka E., Łasocha W., Proniewicz L.M. Influence of sulphur substitution on structural and electrical properties of lithium-manganese spinels. J. Phys. Chem. Solids 2006;67(5-6) 1348-1350.

[11] Chung K.Y., Yoon W.-S., Lee H.S., Yang X.-Q., McBreen J., Deng B.H., Wang X.Q., Yoshio M., Wang R., Gui J., Okada M. Comparative studies between oxygen-deficient $LiMn_2O_4$ and Al-doped $LiMn_2O_4$. Journal of Power Sources 2005;146(1-2) 226-231.

[12] Padhi A.K., Nanjundaswamy K.S., Goodenough J.B. Phospho-olivines as Positive-Electrode Materials for Rechargeable Lithium Batteries. Journal of the Electrochemical Society 1997;144(4) 1188-1194.

[13] Goodenough J.B., Kim Y. Challenges for Rechargeable Li Batteries. Chemistry of Materials 2010;22(3) 587-603.

[14] Molenda J., Molenda M. Composite Cathode Material for Li-Ion Batteries Based on $LiFePO_4$ System. In: Cuppoletti J. (ed.), Metal, Ceramic and Polymeric Composites for Various Uses. InTech; 2011.

[15] Padhi A.K., Nanjundaswamy K.S., Masquelier C., Okada S., Goodenough J.B. Effect of structure on the Fe^{3+}/Fe^{2+} redox couple in iron phosphates. Journal of the Electrochemical Society 1997;144(5) 1609-1613.

[16] Arroyo-de Dompablo M.E., Armand M., Tarascon J.M., Amator U. On-demand design of polyoxianionic cathode materials based on electronegativity correlations: An exploration of the Li_2MSiO_4 system (M = Fe, Mn, Co, Ni). Electrochemistry Communication 2006;8 1292-1298.

[17] Gong Z.L., Li Y.X., Yang Y. Synthesis and electrochemical performance of Li_2CoSiO_4 as cathode material for lithium ion batteries. Journal of Power Sources 2007;174(2) 524-527.

[18] Nyten A., Abouimrane A., Armand M., Gustafsson T., Thomas J.O. Electrochemical performance of Li_2FeSiO_4 as a new Li-battery cathode material. Electrochemistry Communication 2005;7(2) 156-160.

[19] Kokalj A., Dominko R., Mali G., Meden A., Gaberscek M., Jamnik J. Beyond One-Electron Reaction in Li Cathode Materials: Designing $Li_2Mn_xFe_{1-x}SiO_4$. Chemistry of Materials 2007;19(15) 3633-3640.

[20] Molenda M., Dziembaj R., Drozdek M., Podstawka E., Proniewicz L.M. Direct preparation of conductive carbon layer (CCL) on alumina as a model system for direct

preparation of carbon coated particles of the composite Li-ion electrodes. Solid State Ionics 2008; 179(1-6) 197-201.

[21] Sirisopanaporn C., Boulineau A., Hanzel D., Dominko R., Budic B., Armstrong A.R., Bruce P.G., Masquelier C. Crystal Structure of a New Polymorph of Li$_2$FeSiO$_4$. Inorganic Chemistry 2010;49(16) 7446-7451.

[22] Boulineau A., Sirisopanaporn C., Dominko R., Armstrong A.R., Bruce P.G., Masquelier C. Polymorphism and structural defects in Li$_2$FeSiO$_4$. Dalton Transactions 2010;39(27) 6310-6316.

[23] Sirisopanaporn C., Masquelier C., Bruce P.G., Armstrong A.R., Dominko R. Dependence of Li$_2$FeSiO$_4$ Electrochemistry on Structure. Journal of the American Chemical Society 2011;133(5) 1263-1265.

[24] Mali G., Rangusa M., Sirisopanaporn C., Dominko R. Understanding ^{6}Li MAS NMR spectra of Li$_2$MSiO$_4$ materials (M=Mn, Fe, Zn). Solid State Nuclear Magnetic Resonance 2012;42 33–41.

[25] Liu W., Xu Y., Yang R. Synthesis, characterization and electrochemical performance of Li$_2$MnSiO$_4$/C cathode material by solid-state reaction. Journal of Alloys and Compounds 2009;480(2) L1-L4.

[26] Karthikeyan K., Aravindan V., Lee S.B., Jang I.C., Lim H.H., Park G.J., Yoshio M., Lee Y.S. Electrochemical performance of carbon-coated lithium manganese silicate for asymmetric hybrid supercapacitors. Journal of Power Sources 2010;195(11) 3761-3764.

[27] Gummow R.J., Sharma N., Peterson V.K., He Y. Crystal chemistry of the *Pmnb* polymorph of Li$_2$MnSiO$_4$. Journal of Solid State Chemistry 2012;188 32-37.

[28] Gummow R.J., Sharma N., Peterson V.K., He Y. Synthesis, structure, and electrochemical performance of magnesium-substituted lithium manganese orthosilicate cathode materials for lithium-ion batteries. Journal of Power Sources 2012;197 231-237.

[29] Molenda M., Swietoslawski M., Rafalska-Lasocha A., Dziembaj R. Synthesis and Properties of Li$_2$MnSiO$_4$ Composite Cathode Material for Safe Li-ion Batteries. Functional Material Letters 2011;4(2) 135-138.

[30] Dominko R., Bele M., Kokalj A., Gaberscek M., Jamnik J. Li$_2$MnSiO$_4$ as a potential Li-battery cathode material. Journal of Power Sources 2007;174(2) 457-461.

[31] Li Y.X., Gong Z.L., Yang Y. Synthesis and characterization of Li$_2$MnSiO$_4$/C nanocomposite cathode material for lithium ion batteries. Journal of Power Sources 2007;174(2) 528–532.

[32] Aravindan V., Ravi S., Kim W.S., Lee S.Y.,. Lee Y.S. Size controlled synthesis of Li$_2$MnSiO$_4$ nanoparticles: effect of calcination temperature and carbon content for high performance lithium batteries. Journal of Colloid and Interface Science 2011;355(2) 472–477.

[33] Swietoslawski M., Molenda M., Zaitz M., Dziembaj R. C/Li$_2$MnSiO$_4$ as a Composite Cathode Material for Li-ion Batteries, ECS Transaction 2012;41 – *in press*

[34] Molenda M., Dziembaj R., Kochanowski A., Bortel E., Drozdek M., Piwowarska Z. Process for the preparation of conductive carbon layers on powdered supports Int. Patent Appl. No. WO 2010/021557, US Patent Application 20110151112.

Damages and Fractures –
Theoretical and Numerical Modeling

Chapter 5

Biaxial Tensile Strength Characterization of Textile Composite Materials

David Alejandro Arellano Escárpita, Diego Cárdenas, Hugo Elizalde, Ricardo Ramirez and Oliver Probst

Additional information is available at the end of the chapter

1. Introduction

Woven architecture confers textile composites (TC) multidirectional reinforcement while the undulating nature of fibres also provides a certain degree of out-of-plane reinforcement and good impact absorption; furthermore, fibre entanglement provides cohesion to the fabric and makes mould placement an easy task, which is advantageous for reducing production times [1]. These features make TC an attractive alternative for the manufacture of high-performance, lightweight structural components. Another interesting feature of TC is that they can be entangled on a variety of patterns, depending of the specific applications intended. Despite the wide interest of textile composites for industry and structural applications, most of the research efforts for strength characterization has focused on unidirectional composites (UDC), resulting in a large number of failure theories developed for UDC (around 20, as inferred from the conclusions of the World-Wide Failure Excercise (WWFE) [2]); some of the most popular failure models are used indistinctly for UDC and TC: most designers use Maximum Strain, Maximum Stress, Tsai-Hill and Tsai-Wu both for UDC or TC as stated in reference [3], despite the fact that none of the aforementioned failure criteria has been developed specifically for TC, which has led to use of high safety factors in critical structural applications to overcome associated uncertainties [4]. The most successful approaches to predict TC strength are based on phenomenological modeling of interactions between constituents at different scales (matrix-yarn-fiber), providing new insight into TC failure mechanisms. However, the implementation of phenomenological models as design tools is considerably more complex than that of traditional failure criteria, while still exhibiting significant deviation from the scarce experimental data [5] available. This scarcity of experimental data to validate or reject failure theories continues to be a major obstacle for improving TC models. Recent investigations reporting biaxial tensile strength tests in 2D-triaxial TC employing tubular specimens suggested that the failure envelope predicted by the maximum strain criterion fits the experimental data in the tension-tension (T-T) quadrant [6] fairly well. Other tests performed on cruciform specimens indicated that the maximum stress criterion is more adequate [5]; however, the authors of ref. [5] expressed some concerns about the generality of the experimental methodology for the case of non quasi-isotropic lay-up configurations, such as the one studied in their work. In view of the lack of consensus for accurate TC strength prediction [7],[8], and as stated by researchers who participated in the World Wide Failure Exercises (WWFE) [2] more experimental data, better testing methods and properly designed specimens are needed to generate reliable biaxial strength models.

2. Biaxial testing review

Combined multi-axial strength characterization of composites is far from straightforward, as three basic elements are required: i) An apparatus capable of applying multi-axial loads, ii) a specimen capable of generating a homogeneous stress and strain field in a predefined gauge zone, producing failure inside this zone for correct strength characterization, and iii) a measurement system capable of acquiring the applied loads and resulting specimen strains. Although the general procedure is similar to that for uni-axial testing, significant complications arise due to the requirements outlined above; moreover, the required equipment is costly and available generally only at large specialized research centres. Regarding the specimens, the ability of generating a homogeneous multi-axial strain field inside a pre-specified gauge zone is not straightforward mainly due to geometric stress concentrations. Finally, the data acquisition system requires a free surface in order to perform direct measurements. In practice these factors limit the number of combined loads that can be applied to a single specimen to only two, although some researchers have proposed apparatuses designed to apply tri-axial loads, albeit at the expense of limiting the access for full field strain measurements. Efforts on multi-axial testing have been disperse and rather unsuccessful in defining adequate testing methodologies, as evidenced by the lack of standardization by international organisms which have otherwise generated well-known standards for uni-axial characterization of composites, such the ASTM D3039 (standard testing procedures for obtaining tensile properties of polymer matrix composites), British Standard: BS 2782: Part 3: Method 320A-F: Method for obtaining mechanical properties of plastics, BS EN ISO 527 Part 5: Plastics. Determination of tensile properties and test conditions for unidirectional fibre-reinforced plastic composites, CRAG (Composite Research Advisory Group) Test Methods for the Measurement of the Engineering Properties of Fibre Reinforced Plastics, Standard ASTM D6856: Testing procedures for textile composite materials, Japanese Industrial Standard JIS K7054: Tensile Test Method for Plastics Reinforced by Glass Fibre, Russian Standards GOST 25.601-80: Design Calculation and Strength Testing Methods of mechanical testing of polymeric composite materials. In brief, there exist at least seven standards for tensile uni-axial characterization, while none specific standard for bi-axial testing. This demonstrates the need for developing biaxial testing methodologies. In this chapter a review is presented of the state of the art of multi-axial testing with emphasis on biaxial tensile specimens, testing machines and data measurement systems. The reasons for concentrating on biaxial loads are: i) The complexity of testing systems increases considerably with the the number of independent applied. ii) Most structural applications of composites uses thin skins, resulting in shell structures, in which the thickness of the laminates is significantly smaller than the other dimensions. One characteristic of shell structures is that buckling failure modes are the limiting factors in the case compressive loads [9]; consequently, the structural strength depends little on the materials strength and mostly on the geometry and stiffness. On the other hand, when tensile loads are applied to shell, the structures tend to be stable, and the final failure does depend on the materials strength. Evidently, given these fundamentally different failure modes in the cases of compressive and tensile loads, respectively, a combination of biaxial load conditions (compressive-compressive, compressive-tensile, tensile-compressive, and tensile-tensile) can lead to a quite complex behaviour and the need for developing predictive failure models that can account for this complexity.

2.1. Biaxial and multiaxial specimens

To generate useful strength data, a biaxial specimen must be capable of meeting a set of requirements [10],[11],[12],[13]: i) A sufficiently wide homogeneous biaxially-stressed zone must be generated for strain measurements, ii) Failure must occur within this zone. iii) No spurious loads (other than tension/compression) should be acting on the gauge specimen. iv) The specimen should accept arbitrary biaxial load ratios. The very design of specimens that recreate biaxially loaded components has become a constantly evolving field, aiming to provide optimal geometry, manufacture and general arrangement for a valid and reliable test [14]. Specimens designed for biaxial testing can be classified into three main groups: i) tubes, ii) thin plates and iii) cruciforms. A review of these groups and their main features is given below.

2.1.1. Tubular specimens

Multi-axial stress states were formerly created with thin-walled tubes subjected to internal pressure, torsion and axial loads [10],[11],[15]. These specimens allow simultaneous application of tensile and compressive longitudinal loads, as well as tangential and shear loads, therefore representing a versatile scheme for the conduction of multi-axial characterization (Figure 1).

However, the existence of stress gradients across the tubular wall makes this method less accurate than setups based on flat plates, which are also more representative of common industrial applications than the tubular geometry. Some studies also reveal high stress concentrations on the gripping ends. A further disadvantage is a pressure leakage after the onset of matrix failure, although some correction can be provided by internal linings [15].

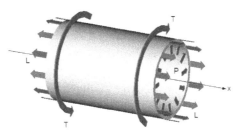

Figure 1. Thin-walled tube specimen.

2.1.2. Thin plates

Round or elliptical flat sheets subject to pressure in the hydraulic bulge test [16], as shown in Figure 2, can develop a biaxial stress state, although the technique has several disadvantages, for example, non-homogeneous stress distributions induced by gripping of the edges [17]. Also, just like the rhomboidal plate case, the loading ratio is shape-dependant [18] and can therefore not be varied during the test to obtain a full characterization.

Figure 2. Elliptical flat sheet used in the bulge test.

2.1.3. Cruciform specimens

Testing biaxially-loaded cruciform specimens represent a more direct approach for obtaining true biaxial stress states, and consequently this method has gained wide acceptance [7],[8],[10],[11],[15]. As suggested by many researchers in the field [7],[10],[13], an ideal cruciform specimen should accomplish the following features: *i*) It should be capable of generating a sufficiently wide and homogenous biaxial stress/strain field in the gauge area, *ii*) failure must occur in the predefined gauge zone, *iii*) the cruciform should accept arbitrary biaxial load ratios for generating a complete failure envelope (within a desired range), *iv*) both the tested and the reinforcement layers should be of the same material, *v*) the transition between the gauge zone and the reinforced regions should be gradual enough as to avoid undesirable high stress concentrations, *vi*) the cruciform fillet radius should be as small as possible in order to reduce stress coupling effects, and *vii*) stress measurements in the test area should be comparable to nominal values obtained by dividing each applied load by its corresponding cross-sectional area. Although various cruciform geometries containing a central-square thinned gauge zone have been proposed in the literature, none can claim full satisfaction of the above requirements due to difficulties inherent to biaxial tests [10]. A cruciform with a thinned central region and a series of limbs separated by slots is presented in Figure 3a. [19]; the slotted configuration allows greater deformations to occur in the thinned section, thus enforcing failure there. Nevertheless, thickness-change can induce undesirable stress concentrations that usually lead to premature failure outside the gauge zone. Also, the extensive machining required for thinning is an undesirable feature.

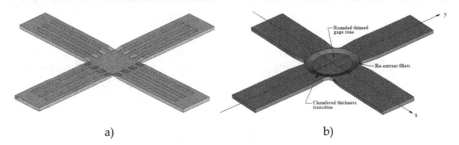

a)   b)

Figure 3. a). Slotted configuration[19] b). Thinned circular zone in the gauge zone [13].

Another cruciform, shown in Figure 3b, with a thinned circular zone in the gauge area [13] exhibits failure outside it, mainly because manufacturing defects caused unexpected higher strength in one axis. The implementation of a rhomboidal shaped test zone is suggested in [20], although, to the authors knowledge, no results with this geometry have been reported so far. Some experiments concluded that loading must be orthogonal to the fibre orientation to produce failure in the test zone [12]. The main difficulty in obtaining an optimal configuration is eliminating stress concentrations in the arms joints. To solve this, an iterative optimization process (numerical/experimental) yielded optimum geometric parameters of the specimen [21]. Results from this study led to a configuration characterized by a thinned square test zone and filleted corners between arms. Given that failure is prone to occur in the arms, reference [23] presented a design where a small cruciform slot is placed in the centre to cause load transfer from the arms to this region (figure 4a).

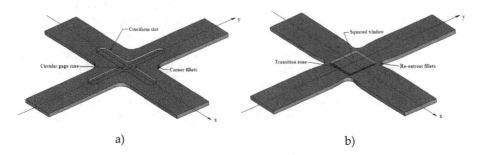

a)   b)

Figure 4. a). Inner cruciform slot [23]. b). Cruciform with thinned rounded square gauge zone and filleted corners [11].

Nevertheless the gauge zone is much reduced, and this makes this specimen useless for TC characterization. In the cruciform proposed by Ebrahim et al [10] failure in the gauge zone is achieved. The design is characterized by a thinned rounded square gauge zone and considers a gradual thickness reduction in the biaxially loaded zone, and also filleted corners as shown in Figure 4b. Results were satisfactory, but it was found that the top and bottom edges of the depression presented high strain gradients. Based on the aforementioned references, a comprehensive study was conducted by the authors to obtain an improved cruciform design. A main feature of this new design is a rhomboid-shaped gauge zone which led to a much more homogeneous strain/strain distribution because of the alleviation of stress concentrations which occur in other designs due to the short distance between the gauge zone and the corners of the arms. Additionally, the corners are filleted to avoid another zone of stress concentration. The specimen is comprised of different layers where the inner layer is under study, whereas the outer ones (equal quantity on each side) are only for reinforcement.

2.1.4. Enhanced rhomboid-windowed cruciform specimen

In order to avoid premature failure due stress concentrations, a modified cruciform was proposed by considering this design concepts: i) Given that fillets are prime examples of stress concentrators, both the cruciform and gauge zone fillets should be as far apart as possible from each other, thus favouring a rhomboid-windowed gauge zone. This modification also intends to minimize regions of stress interactions, which cause lack of homogeneity in the strain field and even premature failure, as reported for some square-windowed specimens [22],[24]. Traditional (instead of re-entrant) fillets were preferred to maintain this stress concentrator as separated as possible from the gauge zone. ii) Since the focus of this research are textile composites (TC), the proposed specimen also features wider arms and a larger gauge zone, seeking to reduce the textile unit cell vs. gauge zone length ratio. This modification is in tune with ASTM standards on testing procedures for textile composites [25]. iii) To avoid polluting the obtained strength data with in-situ effects, adhesion between adjacent layers and other multilayer-related uncertainties, characterization is performed for a single-layer central gauge zone, while a number of reinforcement layers are added outside the gauge zone to enforce failure inside it. The resulting rhomboid windowed cruciform shape was similar to other specimens employed for fatigue characterization of ABS plastic, which report a smooth biaxial strain field at the gauge zone [26]. Basic dimensions were selected from a specimen reported in literature [27]: arm width w = 50mm and cruciform fillets R=25mm. The rhomboid window length l was set identical to the arm width, l=50mm while the window's fillet radius r was set as 10% of l; the geometry is sketched in figure 5. Finite-element (FE) analysis demonstrated that this geometry generates a more uniform strain distribution, while the maximum shear strain in the cruciform fillet is relatively slow.

Once the suitability of a rhomboid windowed cruciform specimen for creating a biaxial strain state was established, a geometrical optimization process based on the experiment design methodology was conducted. Suitable objective functions were defined in order to homogenize the ε_x and ε_y strain fields inside the rhomboid gauge zone while maintaining shear strain γ_{xy} field close to zero. Details of the optimization process can be found in reference [29]

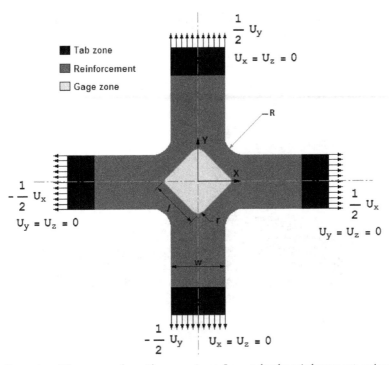

Figure 5. Geometry of the proposed cruciform specimen. Lay up for the reinforcement region is [0]₅, while for the gauge zone is [0] (that is, a single layer). All dimensions are given in mm.

E_{11}	[GPa]	25.0
E_{22}	[GPa]	25.0
V_{12}	[-]	0.2
G_{12}	[GPa]	4.0

Table 1. In-plane measured mechanical properties for a TC conformed of: Epoxy West System 105/206 reinforced with fibreglass cloth style #7520, bidirectional plain weave 8.5 oz./sq. yd, with 18L x 18W threads per inch count.

Evaluation of specimen using finite element analysis was carried by applying boundary conditions as defined in figure 5, with U_x and U_y chosen to produce a maximum strain (ε_x or ε_y) of 2% inside the gauge zone, corresponding to typical failure strain values reported for glass-epoxy TC [5]. The materials properties correspond to a generic plain weave bidirectional textile, as presented in Table 1.

The optimized geometry is defined in Figure 6, while the results of the FE analysis are shown in Figure 7, which splits the geometry into top and bottom sections for simultaneously illustrating the ε_x and γ_x strain fields, respectively, in a single graph; due to full symmetry, the ε_y strain field is identical to the ε_x field when rotated by 90°. The resulting geometry generates a very homogeneous strain field in the gauge zone and keeps shear strains near zero, while keeping shear strains in the fillet regions below the failure value.

These results are believed to represent a great improvement if compared with other specimens reported in the literature.

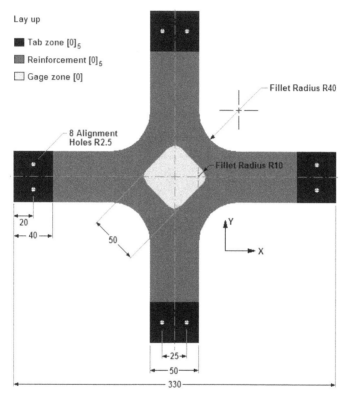

Figure 6. Optimized specimen specifications Dimensions are in mm.

2.2. Biaxial testing machines

To apply biaxial loads on cruciform specimens a specific device is required, which can meet the following requirements [28],[30],[12],[31]: i) The loads applied to a cruciform specimen must be strictly in tension or compression, avoiding spurious shear or bending loads. ii)The restriction previously stated implies that orthogonality among load axes must be guaranteed at all times during the test, and, consequently iii) the centre of the specimen must remain either still or the load axes must displace with it. An efficient method to ensure the previous condition is to apply equal displacements in the loaded axis. These requirements can be accomplished by using an active control system, or by passive mechanical methods, such that the one described later. A review of the most common biaxial testing systems is presented next.

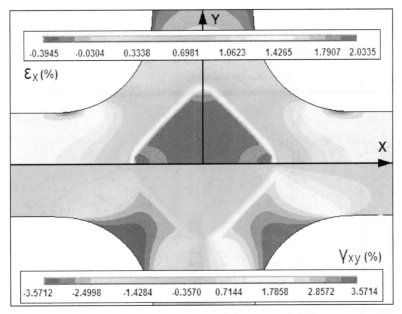

Figure 7. Linear strain field (upper part of the graph) and shear strain field (lower part) within and near the gauge zone of the optimized specimen.

2.2.1. Hydraulic systems

Hydraulic systems rely on hydraulic actuators to apply loads to the specimen; they typically use double-acting pistons with a closed-loop servo control system which sense displacements and/or loads as feedback, as implemented in the design by Pascoe and de Villiers [32]. This configuration (sketched on figure 8) which comprises the use of independent actuator for each applied load, allows the centre of the specimen to move during the test, which is an undesirable condition; this adverse feature can be avoided by implementing a control system that ensures synchronization of opposite actuators [33],[18] thereby avoiding motion of the centre of the specimen.

This configuration also allows the load ratios to be varied in order to obtain a full failure envelope. None of the systems mentioned could ensure equal displacement in both extremes of each axis, even the one using synchronization control, therefore allowing the centre of the specimen to move. If systems are implemented to correct this problem, the design and manufacturing costs inevitably increase. Fessler [34] proposed a machine in which motion is allowed only in one direction at one arm for each cruciform axis. This is the most common basic configuration found in the literature related to biaxial characterization of composites [35],[33]. In an attempt to simplify the previous concept while maintaining symmetric load conditions, some modifications have been proposed; for example, each loading axis, consisting of a pair of opposite hydraulic actuators, can be connected to a common hydraulic line so the force exerted by each side is the same and thus movements of the

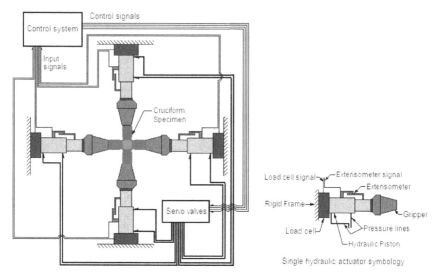

Figure 8. Use of independent actuator per load applied.

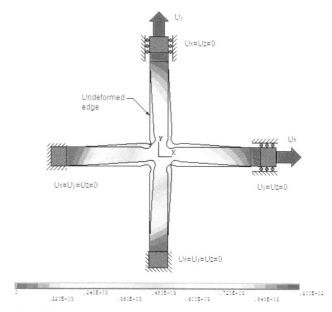

Figure 9. Contour plots of the magnitude of the displacement vector for the case of a configuration where one end of each axis is fixed and the other is displaced.

centre of the specimen are eliminated. Although the common hydraulic line ensures equal force in both extremes of one axis this does not ensure equal displacements. Another variation to hydraulic systems is described in the US Patent No. 5279166, which describes a biaxial testing machine consisting of two independently orthogonal loading axes capable of applying tension and/or compression loads; two ends of the specimen are gripped to fixed ends while the complementary ends are fixed to grips attached to actuators that apply the load, made in an attempt to reduce the complexity and hence the costs of biaxial testing machines (fig. 9). This configuration results in significant displacements of the centre of the specimen, although it is stated that the machine has a mechanism that helps maintain the centre of the specimen and ensure that the loads are always orthogonal. In spite of these features, under large displacements the mechanism used is not capable of maintaining the orthogonally of the loads as shown by a quick finite-element evaluation, whose results are shown Figure 8; moreover, the resulting displacement field is completely asymmetric, a condition which generates undesirable shear stress. While most of the biaxial testing hydraulic machines are original developments, a commercial biaxial testing machine has been developed by the company MTS in conjunction with NASA. It uses four independent hydraulic actuators, each with a load cell and hydraulic grippers, and an active alignment system for the specimen. While solving most of the problems mentioned above, the cost of this system is too high for entry-level composites development laboratories.

2.2.2. Mechanical systems

Mechanical systems owe their name to the fact that they are based on the kinematics of their mechanisms to maintain load symmetry, no matter if the actuators are hydraulic or mechanic; even the application of deadweight to the specimen through systems of ropes, pulleys, levers and bearings has been considered, as presented by Hayhurst et al [36]. In practice, the mechanical systems proposed for the characterization of composite materials are mainly test rigs designed to be adapted to conventional uniaxial testing machines; basically, they are mechanisms consisting of coupled jointed-arms capable of applying in-plane biaxial loads to cruciform specimens. The load ratio is dependent on the geometrical configuration of the device [31] and can therefore be varied only by changing the length of one element, an impractical solution. Similar devices are found in French Patent No. 2579327 [37] and US Patent No. 7204160. A simpler mechanism is presented in US Patent No. 5905205 [30] which uses a four-bar rhomboid-shaped mechanism on which the loading ratios are changed before the test by certain variations in the assembly of the members. One of most practical mechanical systems found consist of four arms, joined at one side to a common block fixed via revolute joints to an universal test machine actuator through a load cell [26] which permits monitoring the applied force, while the other sides are linked, also with revolute joints, to a sliding block each; those blocks slide over a flat plate, fixed to the universal machine's frame. The sliding blocks assemble the grippers which hold the specimen.

2.2.3. A novel biaxial testing apparatus

After reviewing the existing machines and mechanisms on which biaxial tests can be carried out, some conclusions can be drawn; in the case of some of them, the lack of a mechanism that automatically corrects any load difference that could lead to the displacement of the centre of the specimen makes them unsuitable for reliable tests ; in those case where such mechanism does exist, it is controlled by means of an active system that increases design complexity and costs.

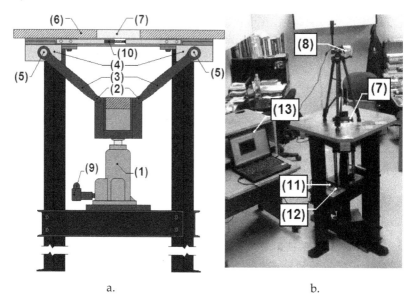

a.

b.

Figure 10. a). Sketch of the biaxial testing machine, showing one load/displacement axis sketch. b). System general arrangement.

The design proposed by the authors considers these drawbacks, as well as the testing requirements previously stated; in addition, construction costs for the novel proposed design are considerably lower compared to other systems. The resulting apparatus is sketched in Figure 10a and a photograph of the completed machine in Figure 10b. The operational principles are described in the following (numbers in parentheses refer to the components identified in the corresponding figures): The loads are applied through a symmetrical slider-crank-slider mechanism meeting the following requirements: The hydraulic piston (1), which is the first slider, is attached on its base to the machine frame and provides the load drive, while its piston is linked by a revolute joint (2) to a pair of arms (3) arranged symmetrically, which in turn are connected by cylindrical joints (5) to the blocks where the grips holding the tab zone of the specimen are installed (4); these cylindrical joints allow to absorb small misalignment in the loads, as established in *iv*.

The grip blocks are lubricated and slide on the lower side of a flat plate (6) featuring a rectangular window (7) allowing a full-field view from the top of the machine, where a high-definition digital camera (HDDC) was installed (8), thereby satisfying the requirement *iii*. A similar arrangement was installed at a right angle with respect the first one, ensuring the independence of the load axes as required by *i*. Data acquisition is conducted by measuring the pressure in the hydraulic cylinders (9) and correcting this information by considering the geometry of the mechanism, while the displacements are measured directly at the grips through resistive displacement sensors (10); all sensors are powered by a power board to provide a common voltage reference (11), and the signals are acquired through a National Instruments 8-channel analogical data acquisition board (12). The information was stored and processed on a laptop (13) by using a Lab View routine.

2.3. Data acquisition techniques

Unlike uniaxial tensile tests in which ultimate failure stress and strains data can be straightforwardly obtained from the collected load and displacement data, in the case of biaxial tests the strength values cannot be calculated directly in this way because the stress and strain fields are not necessarily homogeneous along the specimen and generally depend on the load in a non-trivial way due the complex geometry. For this reason biaxial testing requires a method capable of measuring the full strain field in the biaxially loaded zone of the specimen. Given that strain cannot be measured directly it is necessary to measure the displacement field, from which the strain field can be easily calculated. Using the strain field and a constitutive model the stress field can also be calculated. However, full-field measurement techniques are not standard data acquisition methods and in order to identify the most suitable technique for this research a survey was realized.

2.3.1. Full field strain measurement methods

The first method considered was reflective photo-elasticity; it is based on birefringence, a physical property which consists of the change of the refraction index of a material when shear stresses are applied. It has been used since decade of the 1950s [38], so is a well characterized technique. However, some limiting factors have been identified for the purposes of the curent project: 1) The preparation of the samples is extremely laborious and requires the application of a layer of birefringent material on the surface to be observed, with a thickness of a few millimetres [38].If compared with the thickness of the composite layer under study which is of the orderof about 0.2-0.3mm it is clear that the application of the measurement layer significantly affects the test results. Another technique considered was Moiré interferometry. This technique requires the printing of a pattern of lines on a transparent medium, which is then illuminated by a LASER source, generating an interference pattern which depends on the deformation of the specimen [39]. However, this method has the disadvantage that the data reduction process is tedious and complex [40], and the results heavily depend on the analyst's experience. After considering these options, a technique called digital image correlation (CDI) was identified from biaxial testing literature [10], [11], [41]. The basic concept consists of obtaining digital images of studied geometry on it initial, non-deformed state and after being subjected to a deformation. The surface of the part under study is pre-printed with a random speckle pattern, so that the displacements between corresponding points on photographs of the non-deformed and deformed states, respectively, can be identified by a computer algorithm. This method has some advantage over the ones mentioned above [42]:

i) The experimental setup and specimen preparation are relatively simple; only one fixed CCD camera is needed to record the digital images of the test specimen surface before and after deformation. ii) Low requirements as to the measurement environment: 2D DIC does not require a laser source. A white light source or natural light can be used for illumination during loading. Thus, it is suitable for both laboratory and field applications. iii) Wide range of measurement sensitivity and resolution: Since the 2D DIC method deals with digital images, the digital images recorded by various high spatial-resolution digital image acquisition devices can be directly processed by the 2D DIC method. For the reasons stated above the 2D DIC method is currently one of the most actively used optical measurement techniques and demonstrates increasingly broad application prospects. Nevertheless, the 2D DIC method also has some disadvantages: i) The surface of the planar test object must have a random grey intensity distribution. ii)The measurements depend heavily on the quality of the imaging system. iii) At present, the strain measurement accuracy of the 2D DIC method is lower than that of interferometric techniques, and is not recommended for the measurement of very small and non-homogeneous deformations. Despite these restrictions the low cost associated with equipment and the low specimen preparation requirements makes Digital Image Correlation the preferred technique for the purposes of this study. The drawbacks can be largely avoided by using the highest-definition camcorder commercially available, using an established Digital Image Correlation program and using a specimen that generates a relatively homogeneous strain field. It was shown above that by proper design and optimization a very homogeneous strain field can indeed be obtained in the gauge zone, so this restriction of the DIC technique was of no concern to this project. Finally, the expected strain values were large enough to be safely detected by the DIC technique.

3. Experimental setup

3.1. Specimen manufacture

As stated by recent research [11] the milling process typically employed to thin the gauge zone produces undesirable damage and stress concentrations in unidirectional (UD) composites; for textile composites (TC), milling would exacerbate this problem due to its more complex 3D structure, making milling an unacceptable choice. The main concern is to preserve the integrity of the textile structure, especially when characterizing a single lamina. To generate a damage-free cruciform specimen with a single-layered gauge zone, a novel manufacturing process was developed by the authors, explained below: **1).** Non-impregnated fabric sheets were fixed to a 6mm thick plywood base to ensure dimensional stability, with a printed grid to help proper fibre alignment of each cloth. The whole arrangement was cut into a square pre-form using a water jet, also cutting away the rhomboidal window corresponding to the gauge zone, as shown in the Figure 11. Afterwards, the material was oven-dried at 60°C during 12 hours to eliminate moisture.

2) The following numeric values inside brackets refer to indications given in Figure 12. Two reinforcement layers (1) corresponding to the bottom side of the cruciform specimen were placed in a lamination frame, consisting of a flat surface (2) surrounded by a square border

(3) with a side length equal to that of the specimen. A pre-formed 2-layer rhomboid step (4) was located at the centre, corresponding to the location of the gauge zone, to ensure planarity of the central layer (5). The reinforcement layers were manually resin-impregnated and, immediately after this, the central layer (5) was placed and impregnated. Finally, the process was repeated for the last two reinforcement layers (6), as shown. Room environment was controlled during the lamination process at a temperature of 80±2°C and 50-60% relative humidity. Immediately after the impregnation process was completed, the laminate was placed in a vacuum bag consisting of a peel ply (7), perforated film (8), bleeder cloth (9) and the bag itself (10), using sealing tape to ensure vacuum seal (11). 0.8 bar vacuum pressure was applied through a valve located at a corner (12), sufficiently away from where the final shape would be cut. The whole arrangement was cured during 4 hours inside a pre-heated oven at 80±2°C, as measured by a thermo-couple (13) located at the gauge zone, as shown in Figure 12. **3)** After curing, the final cruciform geometry was obtained through water jet cutting. Nine specimens were prepared meeting the dimensional specifications in Figure 6.

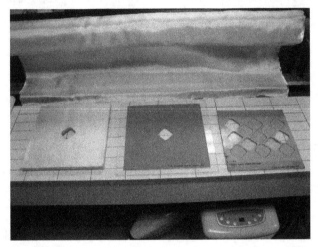

Figure 11. Rhomboid window cutted on the reinforcement layers and other auxiliary tools.

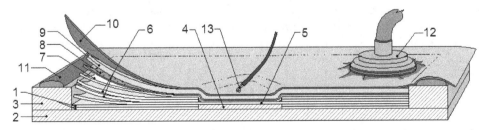

Figure 12. Arrangement for specimen manufacture.

To provide visual reference for the digital image correlation (DIC) strain field measurement [24][41], specimens were painted with a black-dot random speckle pattern over a white-mate primer, as shown in Figure 13. This technique was preferred over the spraying technique reported in [41], as it might result in an inadequate control of the dot size distribution, leading to uncertainty in the DIC measurements. Additionally, five uniaxial, [0]₅ layup specimens were prepared in order to perform uniaxial tests to provide precise input data for the development of failure criteria.

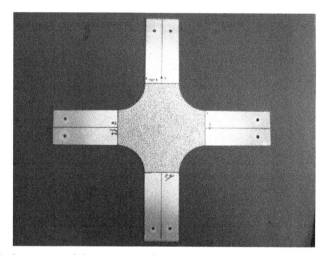

Figure 13. Finished specimen exhibiting its speckle pattern used with the digital image correlation technique

3.2. Biaxial testing

Experimental validation of the optimized cruciform was conducted with the biaxial testing apparatus described in section 2.2.3 as follows: after mounting the specimen in the grippers, a pre-load of 500N was applied to each axis prior to tightening the mounting bolts. (Figure 14) Then, preload and alignment bolts were removed, setting the measured displacements and loads to zero. A high-quality video of the specimen was recorded with a high definition cam coder with adjustable focus and exposure parameters functions for subsequent DIC analyses.

A chronometer synchronized with the computer clock was placed near the specimen and inside the camera vision field, to ensure its inclusion in the captured images; this provided a time reference to relate each video frame with correspondent load data. After starting video recording and the data capture routine, biaxial displacements were applied at a rate of 1mm/min until final failure. This load rate was selected based on the ASTM 3039 standard [43], which recommends a displacement speed such that failure occurs 1 to 10 minutes after the start of the test.

Data acquisition and reduction was conducted as follows: two video frames were taken from the recorded video sequence, one corresponding to the beginning of the test and another just prior to final failure, as shown in figure 15. Both images were fed into the open access software DIC2D (developed by Dr. Wang's team at the Catholic University of America) to obtain the full strain field (ε_x, ε_y and γ_{xy}). The three tests performed covered a range of biaxial ratio BR values in the vicinity of the critical condition BR=1: BR=1.5 (Test #1), BR=1.25 (Test #3), and BR=1 (Test #5). Figure 15 shows the final failure sequence representative of the tests conducted. It should be noted that the failure occurred well within the gauge zone as expected from the FE-predicted strain fields.

Figure 14. Cruciform specimen mounted in the biaxial testing machine.

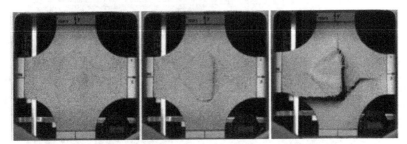

Figure 15. Final failure sequence recorded at 30 frames per second.

The final failure is clearly fibre-dominated, due to its catastrophic nature; it is possible to identify the final failure onset region inside the rhomboidal gauge zone, as required for a successful test. Regarding the strain field, it can be seen from Figure 16 that the agreement between the experimentally results (obtained from DIC) and the FE prediction is remarkably good. The DIC and the FE images show the same symmetry of the experimental shear strain pattern and similar homogeneity and smoothness, and the absolute strain values cover a similar range. This can be considered an additional indicator of the success of the experimental procedure presented in this work.

The same procedure used to characterize the ultimate strain can be used to obtain the matrix onset failure envelope (as opposed to fibre failure), but due to the fact that this phenomenon cannot be deduced visually a different approach was used for this purpose. The load vs. displacement plots were used to identify the change in the slope which evidences matrix damage, as shown in the Figure 17. This method is proposed as an extrapolation of the method employed for uniaxial tests defined by the ASTM 3039 standard for the uniaxial tensile characterization of composites [43]. Linear fits were obtained for every linear segment of curves corresponding to every perpendicular axis, and the intersections were calculated solving the resulting equations, which allowed to quantify the strain values corresponding to the onset of matrix damage, considering that the latter occurs at the first observed slope change. Once the displacement and strain were identified, the digital image corresponding were used to perform a DIC analysis and to get the full field strain in the same fashion described previously.

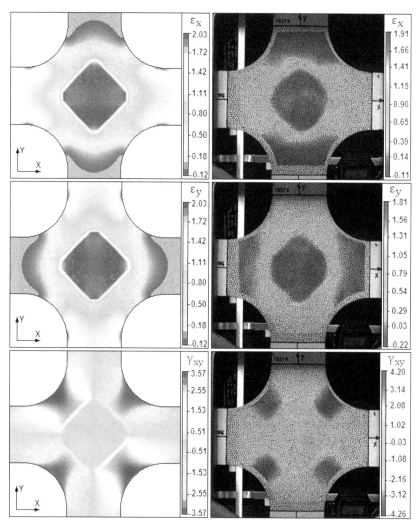

Figure 16. FE vs DIC strain field comparison for Test #5. The first column corresponds to the FE results, while the second column exhibits the results of the digital image correlation (DIC) process. The first and second row show the linear strain field, while the last row exhibits the shear strain stress field.

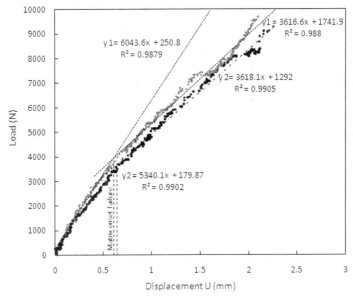

Figure 17. Load vs diplacement for biaxial test #3. The location where the change of slope occurs is interpreted as the onset of matrix failure.

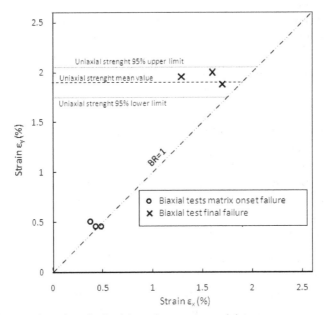

Figure 18. Failure envelope data obtained from the experimental data

Is important to remark that the use of the slope change in the load vs. displacements curves can be significantly influenced by geometrical effects and materials non-linearity, and other auxiliary techniques such as sonic emission or in-situ x-ray scanning should be employed to verify that this change can be effectively used as a matrix damage onset indication.

Experimental strength data for single layer biaxial strength obtained from the experimental program are presented in Figure 18 as well as data from uniaxial test performed on five layer specimens. Confidence intervals calculated for the failure strains observed on uniaxial tests are presented in the figure. It should be noted that single-layer strength data fall inside the 95% confidence limits which suggest that interactions between ε_1 and ε_2 strains are significant for single layer laminates. This finding should not be used as design criterion before more experimental data are obtained, but it gives a good indication of the feasibility of the methodology presented for the purposes of failure analysis.

4. Conclusions

Improvement over existing cruciform specimens for biaxial testing was achieved by proposing a specimen with rhomboidal thinned gauge zone, based on conclusions from a qualitative stress concentration analysis. An optimization based on the experiment design methodology was performed to achieve a highly homogeneous strain distribution withinu the rhomboidal gage zone while shear strains in the cruciform fillets were kept well below the failure values in order to avoid the premature failure typically affecting this kind of specimens. The resulting geometry generates very homogeneous strain field within the gauge zone and keeps shear strains near zero, while keeping shear strains in fillets below the failure value; this is believed to represent a great improvement over other specimens reported in the literature. In addition to meeting the requirements for equi-biaxial tests, the specimen was evaluated under various biaxial ratios, demonstrating that is practically insensitive to biaxial ratio, and hence can be used without any modification to obtain the full tension-tension failure envelope.

A manufacturing process which avoids machining operations normally required to generate the thinned gauge zone was developed, in an attempt to preserve the textile architecture from machining micro-damage. It consists in cutting the rhomboidal windows from the reinforcement layers prior to its matrix impregnation by using a water jet cutting machine. Despite the highly manual work involved in the specimens manufacturing process, it measured specifications were according to those extrapolated from ASTM 3039 for composite materials unidirectional samples.

Validation of the specimen's geometry and manufacturing technique was made through experimental testing, which were conducted on the in-house-developed biaxial machine. The cruciform's full strain field was measured via digital image correlation; the results demonstrate, in close agreement with the results obtained from finite-element(FE) simulations, that the specimen generates a significantly more homogeneous biaxial load state in the gauge zone than others reported in literature, and failure occurs, for all the tests, inside the gauge zone, as intended.

Author details

David Alejandro Arellano Escárpita*, Diego Cárdenas,
Hugo Elizalde and Ricardo Ramirez
*Mechatronics Engineering Department, Instituto Tecnológico y de Estudios
Superiores de Monterrey, Campus Ciudad de México, Col. Ejidos de Huipulco, Tlalpan,
México D.F., Mexico*

Oliver Probst
*Physics Department, Instituto Tecnológico y de Estudios Superiores de Monterrey,
Campus Monterrey, Monterrey, N.L., Mexico*

5. References

[1] Crookston JJ, Long AC, Jones IA. A summary review of mechanical properties prediction methods for textile reinforced polymer composites. Proceedings of the Institution of Mechanical Engineers, Part L: Journal of Materials: Design and Applications 2005; 219(2):91-109.

[2] Soden PD, Kaddour AS, Hinton M.J. Recommendations for designers and researchers resulting from the world-wide failure exercise. Composites Science and Technology 2004; 64(3-4): 589–604.

[3] Kaddour AS, Hinton MJ. Instructions To Contributors Of The Second World-Wide Failure Exercise (Wwfe-II): Part(A).

[4] Lomov SV, Huysmans G, Luo Y, Parmas RS, Prodromou A, Verpoest I, Phelan FR. Textile composites: modelling strategies. Composites: Part A 2001; 32(10): 1379-1394.

[5] Welsh JS, Mayes JS, Key CT, McLaughlin RN. Comparison of MCT failure prediction techniques and experimental verification for biaxially loaded glass fabric-reinforced composite laminates. Journal of Composite Materials 2004; 38(24):2165-2181.

[6] Swanson SR, Smith LV. Comparison of the biaxial strength properties of braided and laminated carbon fibre composites. Composites: Part B 1996; 27(1):71-77.

[7] Welsh JS, Adams DF. An experimental investigation of the biaxial strength of IM6/3501-6 carbon/epoxy cross-ply laminates using cruciform specimens. Composites: Part A 2002; 33(6):829-839.

[8] Welsh JS, Mayes JS, Biskner A. 2-D biaxial testing and failure predictions of IM7/977-2 carbon/epoxy quasi-isotropic laminates. Composite Structures 2006; 75(1-4):60-66.

[9] Ng RK, Yousefpour A, Uyema M, Ghasemi MN. Design, Analysis, Manufacture, and Test of Shallow Water Pressure Vessels Using E-Glass/Epoxy Woven Composite Material for a Semi-Autonomous Underwater Vehicle. Journal of Composite Materials 2002; 36(21): 2443-2478.

---

* Corresponding Author

[10] Ebrahim L, Van Paepegem W, Degrieck J, Ramault C, Makris A, Van Hemelrijck D. Strain distribution in cruciform specimens subjected to biaxial loading conditions. Part 1: Two-dimensional versus three-dimensional finite element model. Polymer Testing 2010; 29(1):7-13.

[11] Ebrahim L, Van Paepegem W, Degrieck J, Ramault C, Makris A, Van Hemelrijck D. Strain distribution in cruciform specimens subjected to biaxial loading conditions. Part 2: Influence of geometrical discontinuities. Polymer Testing 2010; 29(1):132-138.

[12] Fawaz, Z. Étude analytique, numérique et expérimentale portant sur la rupture et la fatigue biaxiales des lamelles renforcées de fibres. PhD Thesis, Sherbrooke: Université de Sherbrooke, 1992.

[13] Gupte AA. Optimization of Cruciform Biaxial Composite Specimen. In: Master on Science Thesis. South Dakota State University, 2003.

[14] Sepúlveda CG. Biaxial Testing of Composite Materials: Technical Specifications and Experimental Set-up Maestría en Ciencias con Especialidad en Sistemas de Manufactura México Mayo 2009

[15] Soden PD, Hinton MJ, Kaddour AS. Biaxial test results for strength and deformation of a range of E-glass and carbon fibre reinforced composite laminates: failure exercise benchmark data. In: Failure Criteria in Fibre Reinforced Polymer Composites: the World-Wide Failure Exercise. Oxford: Elsevier Science LTD, 2004, p. 52-96.

[16] Bird JE, Duncan J. Strain hardening at high strain in aluminum alloys and its effect on strain localization. Metallurgical and Materials Transactions A 1981; 12(2): 235-241.

[17] Dudderar TF, Koch, Doerries DE. Measurement of the shapes of foil bulge-test samples. Experimental Mechanics 1977; 17(4): 133-140.

[18] Boehler JP, Demmerle S, Koss S. A New Direct Biaxial Testing Machine for Anisotropic Materials. Experimental Mechanics 1944; 34(1): 1-9.

[19] Mönch, E., and D. Galster. "A Method for Producing a Defined Uniform Biaxial Tensile Stress Field." British Journal Of Applied Physics 14, no. 11 (1963).

[20] Sacharuk, Z. Critères de rupture et optimisation des éléments en matériaux composites. PhD Thesis. Université de Sherbrooke. Sherbrooke 1990

[21] Youssef, Y. Résistance des composites stratifiés sous chargement biaxial: validation expérimentale des prédictions théoriques. PhD Thesis, Sherbrooke: Université de Sherbrooke, 1995.

[22] Arellano D, Sepúlveda G, Elizalde H, Ramírez R. "Enhanced cruciform specimen for biaxial testing of fibre-reinforced composites." International Materials Research Congress. Cancun, 2007.

[23] Yong Yu, Min Wan, Xiang-Dong Wu, Xian-Bin Zhou. Design of a cruciform biaxial tensile specimen for limit strain analysis by FEM. Journal of Materials Processing Technology 2002; 123:(67-70)

[24] Antoniou AE, Van Hemelrijck D, Philippidis P. Failure prediction for a glass/epoxy cruciform specimen under static biaxial loading. Composite Science and Technology 2010; 70(8):1232-1241.

[25] ASTM Standard D6856–03, Standard Guide for Testing Fabric-Reinforced Textile Composite Materials, 2003.

[26] Kwon HJ, Jar PYB, Xia Z. Characterization of bi-axial fatigue resistance of polymer plates. Journal of Materials Science 2005; 40(4):965– 972.

[27] Gdoutos EE, Gower M, Shaw R, Mera R. Development of a Cruciform Specimen Geometry for the Characterisation of Biaxial Material Performance for Fibre Reinforced Plastics. Experimental Analysis of Nano and Engineering Materials and Structures. Springer Netherlands, 2007. p. 937-938.

[28] [28] Chaudonneret M, Gilles P, Labourdette R, Policella H. Machine d'essais de traction biaxiale pour essais statiques et dynamiques. La Recherche Aérospatiale 1977:299-305.

[29] David Alejandro Arellano Escárpita. Experimental investigation of textile composites strength subject to biaxial tensile loads. Ph.D. Thesis. Instituto Tecnológico y de Estudios Superiores de Monterrey. Monterrey 2011.

[30] Clay SB. Biaxial Testing Apparatus. United States of America Patent 5905205. 18 May 1999.

[31] Ferron G, Makinde A. Design and Development of a Biaxial Strength Testing Device. Journal of Testing and Evaluation (ASTM) 16, no. 3 (May 1988).

[32] [32] Pascoe KJ, de Villiers JWR. Low Cycle Fatigue of Steels Under Biaxial Straining. Journal of Strain Analysis 1967; 2(2).

[33] Welsh JS, Adams DF. Development of an Electromechanical Triaxial Test Facility for Composite Materials. Experimental Mechanics 2000; 40(3): 312-320.

[34] Fessler H, Musson JK. A 30 Ton Biaxial Tensile Testing Machine. Professional Engineering Publishing 1969; 4(1): 22-26.

[35] Makinde A, Thibodeau L, Neale K. Development of an apparatus for biaxial testing using cruciform specimens. Experimental Mechanics 1992; 32(2): 138-144.

[36] Hayhurst DR. A Biaxial-Tension Creep-Rupture Testing Machine. Professional Engineering Publishing 1973; 8(2): 119-123.

[37] Ferron, G. Dispositif perfectionné d'essais de traction biaxiale. France. 26 Septembre 1986.

[38] Doyle JF, Phillips JWP. Manual on experimental stress analysis. Society for Experimental Mechanics (U.S.). Society for Experimental Mechanics, 1989.

[39] Post D, Han B. Chap 22, "Moire Interferometry," Handbook on Experimental Mechanics, W. N. Sharpe, Jr., ed., Springer-Verlag, NY, 2008.

[40] Gre'diac M. The use of full-field measurement methods in composite material characterization: interest and limitations. Composites: Part A 2004; 35(7-8):751-761.

[41] Lecompte D, Smits A, Bossuyt S, Sol H, Vantomme J, Van Hemelrijck D, Habraken AM. Quality assessment of speckle patterns for digital image correlation. Optics and Lasers in Engineering 2006; 44(11):1132-1145.

[42] Bing P, Kemao Q, Huimin X, Anand A. Two-dimensional digital image correlation for in-plane displacement and strain measurement: a review. Measurement Science and Technology 2009; 20(6): 1-17.

[43] ASTM Standard D 3039/D 3039M, Standard Test Method for Tensile Properties of Polymer Matrix Composite Materials, 2000.

Chapter 6

Finite Element Implementation of Failure and Damage Simulation in Composite Plates

Milan Žmindák and Martin Dudinský

Additional information is available at the end of the chapter

1. Introduction

Composite materials are now common engineering materials used in a wide range of applications. They play an important role in the aviation, aerospace and automotive industry, and are also used in the construction of ships, submarines, nuclear and chemical facilities, etc.

The meaning of the word damage is quite broad in everyday life. In continuum mechanics the term damage is referred to as the reduction of the internal integrity of the material due to the generation, spreading and merging of small cracks, cavities and similar defects. Damage is called elastic, if the material deforms only elastically (in macroscopic level) before the occurrence of damage, as well as during its evolution. This damage model can be used if the ability of the material to deform plastically is low. Fiber-reinforced polymer matrix composites can be considered as such materials.

The use of composite materials in the design of constructions is increasing in traditional structures such as for example development of airplanes or in the automotive industry. Recently this kind of materials is used in development of special technique and rotating systems such as propellers, compressor turbine blades etc. Other applications are in electronics, electrochemical industry, environmental and biomedical engineering (Chung, 2003).

The costs for designing of composite structures is possible partially eliminate by numerical simulation of solving problem. In this case the simulation is not accepted as universal tool for analyzing of systems behaviour but it is an effective alternative to processes of experimental sciences. Simulations support development of new theories and suggestion of new experiments for testing these theories. Experiments are necessary for obtaining of input data into simulation programs and for verification of numerical programs and models.

Laminated composites have a lot of advantages but in some cases they show different limitations that are caused by stress concentrations between layers. Discontinuous change of material properties is reason for occurrence of interlaminar stresses that often cause delamination failure (Zhang & Wang, 2009). Delamination may originate from manufacturing imperfections, cracks produced by fatigue or low velocity impact, stress concentration near geometrical/material discontinuity such as joints and free edges, or due to high interlaminar stresses (Elmarakbi, 2009)

Delaminations in layered plates and beams have been analyzed by using both cohesive damage models and fracture mechanics. Cohesive elements are widely used, in both forms of continuous interface elements and point cohesive elements (Cui, 1993), at the interface between solid finite elements to predict and to understand the damage behaviour in the interfaces of different layers in composite laminates. In the context of the fracture mechanics approach (Sládek, et al. , 2002), it allows us to predict the growth of a preexisting crack or defect. In a homogeneous and isotropic body subjected to a general loading condition, a crack tends to grow by kinking in a direction such a pure mode I condition at its tip is maintained. On the contrary, delaminations in laminated composites are constrained to propagate in its own plane because the toughness of the interface is relatively low in comparison to that of the adjoining material. Therefore a delamination crack propagates with its advancing tip in mixed mode condition and, consequently, requires a fracture criterion including all three mode components.

The theory of crack growth may be developed by using one of two approaches. First, the Griffith energetic (or global) approach introduces the concept of energy release rate (ERR) G as the energy available for fracture on one hand, and the critical surface energy G_r as the energy necessary for fracture on the other hand. Alternatively, the Irwin (local) approach is based on the stress intensity factor concept, which represents the energy stress field in the neighborhood of the crack tip. These two approaches are equivalent and, therefore, the energy criterion may be rewritten in terms of stress intensity factors.

Microcracking in a material is almost always associated with changes in mechanical behavior of the material. The problem of microcracking in fiber-reinforced composites is complicated due to the multitude of different microcracking modes which may initiate and evolve independently or simultaneously. Continuum Damage Mechanics (CDM) considers damaged materials as a continuum, in spite of heterogenity, micro-cavities, and micro-defects and is based on expressing of stiffness reduction caused by damage, by establishing effective damage parameters which represent cumulative degradation of material. There are basically two categories of CDM models used for estimating the constitutive behavior of composite materials containing microcracks –phenomenological models and micromechanics models.

The phenomenological CDM models employ scalar, second order or fourth order tensors using mathematically and thermodynamically consistent formulations of damage mechanics. Damage parameters are identified through macroscopic experiments and in general, they do not explicitly account for damage mechanism in the microstructure. On the other hand the micromechanics-based approaches conduct micromechanical analysis of representative volume element (RVE) with subsequent homogenization to predict evolving material damage behavior (Mishnaevsky, 2007). Most damage models do not account for the evolution of damage or the effect of loading history (Jain & Ghosh, 2009). Significant error can consequently accrue in the solution of problems, especially those that involve nonproportional loading. Some of these homogenization studies have overcome this shortcoming through the introduction of simultaneous RVE-based microscopic and macroscopic analysis in each load step. However, such approaches can be computationally very expensive since detailed micro-mechanical analyses need to be conducted in each load step at every integration point in elements of the macroscopic structure.

Jain & Ghosh (2009) have developed a 3D homogenization-based continuum damage mechanics (HCDM) model for fiber-reinforced composites undergoing micro-mechanical damage. Micromechanical damage in the RVE is explicitly incorporated in the form of fiber-matrix interfacial debonding. The model uses the evolving principal damage coordinate system as its reference in order to represent the anisotropic coefficients, which is necessary for retaining accuracy with nonproportional loading. The HCDM model parameters are calibrated by using homogenized micromechanical solutions for the RVE for a few strain histories.

There are many works written about damage of composite plates and many models for various types of damage, plates or loading have been developed. Shu (2006) presented generalized model for laminated composite plates with interfacial damage. This model deals with three kinds of interfacial debonding conditions: perfect bonding, weak bonding and delamination. Iannucci & Ankersen (2006) described unconventional energy based composite damage model for woven and unidirectional composite materials. This damage model has been implemented into FE codes for shell elements, with regard to tensile, compressive and shear damage failure modes. Riccio & Pietropaoli (2008) dealt with modeling damage propagation in composite plates with embedded delamination under compressive load. The influence of different failure mechanisms on the compressive behavior of delaminated composite plates was assessed, by comparing numerical results obtained with models characterized by different degrees of complexity. Tiberkak et al. (2008) studied damage prediction in composite plates subjected to low velocity impact. Fiber-reinforced composite plates subjected to low velocity impact were studied by use of finite element analysis where Mindlin's plate theory and 9-node Lagrangian element were considered. Clegg et al. (2006) worked out interesting study of hypervelocity impact damage prediction in composites. This study reports on the development of an extended orthotropic continuum material model and associated material characterization techniques for the simulation and validation of impacts onto fiber-reinforced composite materials. The model allows to predict the extent of damage and residual strength of the fiber-reinforced composite material after impact.

Many studies of an effect of various aspects of damage process on behavior of composite plates can be found in the literature (e.g. Gayathri et al. , 2010). Many authors have been dealing with problematic of damage of composite plated under cyclic loading or problematic of impact fatigue damage (e.g. Azouaoui et al. , 2010). Composite materials are becoming more and more used for important structural elements and structures, so the problematic of fatigue damage of composites is becoming more and more actual. Numerical implementation of damage is not simple. Finite element method (FEM) is the most utilized method for modeling damage. Fast Multipole BEM method or various meshless methods are also establishing at the present time.

Firstly, the goal of this chapter is to present the numerical results of the delamination analysis of two laminae with different thickness with two orthotropic material properties and subjected to a pair of opposed forces. For this goal we used commercial FEM software ANSYS and the mode I, II, and III components of energy release rate (ERR) were calculated. Secondly, the goal is to present the numerical results of elastic damage of thin composite plates. The analysis was performed by user own software, created in MATLAB programming language. This software can perform numerical analysis of elastic damage using FEM layered plate finite elements based on the Kirchhoff plate theory.

This chapter is organized in three sections: Second section is focused on failure modeling in laminates by using standard shear deformable elements, whereas interface elements were used for the interface model. The delamination propagation is controlled by the critical ERR. In third section, in first, a general description of damage is provided, then damage model used is examined. Finally, some numerical results obtained for damage of plate are presented.

2. Theoretical background of failure modeling in laminates

The mechanisms that lead to failure in composite materials are not yet fully understood, especially for matrix or fiber compression. Strength-based failure criteria are commonly used with the FEM to predict failure events in composite structures. Numerous continuum-based criteria have been derived to relate internal stresses and experimental measures of material strength to the onset of failure (Dávila et al., 2005). In Fig. 1 a laminate contains a single in-plane delamination crack of area Ω_D with a smooth front $\partial\Omega_D$. The laminate thickness is denoted by h_0. The x-y plane is taken to be the mid-plane of the laminate, and the z-axis is taken positive downwards from the mid-plane.

2.1. Plate finite elements for sublaminate modeling

Each sublaminate is represented by an assembly of first order shear deformable (FSDT) plate elements bonded by zero-thickness interfaces in the transverse direction as shown in Fig. 2. The delamination plane separates the delaminated structure into two sublaminates of thickness h_1, h_2 and each sublaminate consist the upper n_u plates and the lower n_l plates. Each plate element is composed from one or few physical fiber-reinforced plies with their material axes arbitrarily oriented. Lagrangian multipliers through constraint equations (CE) are used for enforcing adhesion between the plates inside each sublaminate. Accordingly, the displacements in the z-th plate element, in terms of a global reference system located at the laminate mid-surface, are expressed by (e.g. Carrera 2002; Reddy 1995)

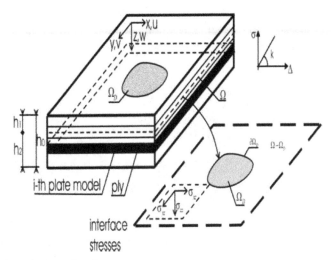

Figure 1. Delaminated composite plate

$$u_i(x,y,z) = u_i^0(x,y) + (z - z_i)\, \psi_{xi}(x,y)$$
$$v_i(x,y,z) = v_i^0(x,y) + (z - z_i)\, \psi_{yi}(x,y) \tag{1}$$
$$w_i(x,y,z) = w_i^0(x,y)$$

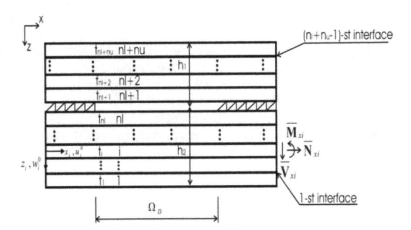

Figure 2. Laminate subdivision in plate elements

where u_i, v_i refer to the in-plane displacements, and w_i to the transverse displacements through the thickness of the i-th plate element, u_i^0, v_i^0, w_i^0, are the displacements at the mid-surface of the i-th plate element and $\psi_{xi}(x,y), \psi_{yi}(x,y)$ denote rotations of transverse normals about y and x, respectively.

At the reference surfaces, the membrane strain vector ε_i, the curvature κ_i, and the transverse shear strain γ_i, respectively are defined as

$$
\left\{ \begin{array}{c} \varepsilon_{xxi} \\ \varepsilon_{yyi} \\ \gamma_{xyi} \end{array} \right\} = \left\{ \begin{array}{c} \dfrac{\partial u_i^0}{\partial x} \\ \dfrac{\partial v_i^0}{\partial y} \\ \dfrac{\partial u_i^0}{\partial y} + \dfrac{\partial v_i^0}{\partial x} \end{array} \right\}, \left\{ \begin{array}{c} \kappa_{xxi} \\ \kappa_{yyi} \\ \kappa_{xyi} \end{array} \right\} = \left\{ \begin{array}{c} \dfrac{\partial \psi_{xi}^0}{\partial x} \\ \dfrac{\partial \psi_{yi}^0}{\partial y} \\ \dfrac{\partial \psi_{xi}^0}{\partial y} + \dfrac{\partial \psi_{yi}^0}{\partial x} \end{array} \right\}, \left\{ \begin{array}{c} \gamma_{yzi} \\ \gamma_{xzi} \end{array} \right\} = \left\{ \begin{array}{c} \psi_{xi}^0 + \dfrac{\partial w_i^0}{\partial y} \\ \psi_{xi}^0 + \dfrac{\partial w_i^0}{\partial x} \end{array} \right\} \quad (2)
$$

The constitutive relations between stress resultants and corresponding strains are given in (Reddy & Miravete, 1995; Žmindák, 2010). In these works standard FSDT finite elements available in ANSYS software are used (ANSYS, 2007) to join these elements at the interfaces inside each sublaminate using CE or rigid links characterized by two nodes and three degrees of freedom at each node.

2.1.1. Interface elements for delamination modeling

Delamination is defined as the fracture of the plane separating two plies of a laminated composite structure (Fenske, et al., 2001). This fracture occurs within the thin resin-rich layer that forms between plies during the manufacturing process. Perfect adhesion is assumed in the undelaminated region $\Omega - \Omega_D$, whereas sub-laminates are free to deflect along the delaminated region Ω_D but not to penetrate each other. A linear interface model, is introduced along $\Omega - \Omega_D$ to enforce adhesion. The constitutive equation of the interface involves two stiffness parameters, k_z, k_{xy}, imposing displacement continuity in the thickness and in-plane directions, respectively, by treating them as penalty parameters. The relationship between the components of the traction vector σ acting at the lower surface of the upper sublaminate, σ_{zx}, σ_{zy} and σ_{zz}, in the out-of-plane (z) and in the in-plane (x and y) directions, respectively, and the corresponding components of relative interface displacement vector Δ, Δu, Δv and Δw is expressed as

$$
\sigma = K \, \Delta
$$

$$(3)$$

Interface elements are implemented using COMBIN14 element. Relative opening and sliding displacements are evaluated as the difference between displacements at the interface between the lower and the upper sublaminate.

2.1.2. Contact formulation for damage interface

In order to avoid interpenetration between delaminated sublaminates in the delaminated region Ω_D, a unilateral frictionless contact interface can be introduced, characterized by a zero stiffness for opening relative displacements ($\Delta w \geq 0$) and a positive stiffness for closing relative displacements ($\Delta w \leq 0$), then the contact stress σ_{zz} is

$$\sigma_{zz} = \frac{1}{2}\left(1 - \text{sign}(\Delta w)\right) k_z \Delta w \tag{4}$$

where k_z is the penalty number imposing contact constraint and sign is the signum function. A very large value for k_z restricts sublaminate overlapping and simulates the contact condition. Unilateral contact conditions may be implemented in ANSYS using COMBIN39. This element is a unidirectional element with nonlinear constitutive relationships with appropriate specialization of the nonlinear constitutive law according to (4).

If we introduce a scalar damage variable D with the value of 1 for no adhesion and the value of 0 for perfect adhesion, we get a single extended interface model with constitutive law valid both for undelaminated $\Omega - \Omega_D$ and delaminated Ω_D areas. Consequently constitutive law can be expressed as

$$\sigma = (1 - D) \mathbf{K} \Delta \tag{5}$$

In this work we use the formulation via FEs, related to plate elements, interface elements and Lagrange multipliers. It is worth noting that in commercial FEA packages the Lagrange multipliers are represented by either CE or rigid links, whereas interface elements are implemented by the analyst using a combination of spring elements (COMBIN14) and CE.

2.1.3. Mixed mode analysis

In order to predict crack propagation in laminates for general loading conditions, ERR distributions along the delamination front are needed. Fracture mechanics assumes that delamination propagation is controlled by the critical ERR. Delamination grows on the region of the delamination front where the following condition is satisfied and is of the form

$$\left(\frac{G_I(s)}{G_I^c}\right)^{\alpha} + \left(\frac{G_{II}(s)}{G_{II}^c}\right)^{\beta} + \left(\frac{G_{III}(s)}{G_{III}^c}\right)^{\gamma} = 1 \tag{6}$$

where α, β and γ are mixed mode fracture parameters determined by fitting experimental test results.

The critical ERR $\left(G_I^c, G_{II}^c, G_{III}^c\right)$ material properties can be evaluated from experimental procedures. The closed-form expressions for the ERR are (Barbero, 2008)

$$G(s) = \frac{1}{2} \lim_{k_z, k_{xy} \to \infty} \left[k_z \Delta w^2(s) + k_{xy} \Delta u^2(s) + k_{xy} \Delta v^2(s) \right], \Delta w(s) \ge 0 \qquad (7)$$

$$G = G_I(s) + G_{II}(s) + G_{III}(s)$$

$$G_I(s) = \begin{cases} \lim_{k_z, k_{xy} \to \infty} \frac{1}{2} k_z \Delta w^2(s) & \text{if} \quad \Delta w(s) \ge 0 \\ 0 & \text{if} \qquad \Delta w(s) < 0 \end{cases}$$

$$G_{II}(s) = \lim_{k_z, k_{xy} \to \infty} \frac{1}{2} k_{xy} \Delta u_n^2(s)$$

$$G_{III}(s) = \lim_{k_z, k_{xy} \to \infty} \frac{1}{2} k_{xy} \Delta u_t^2(s)$$

They are obtained by means of the interface model, using FE code to check whether propagation occurs. Once made a global FEA of the laminate, then the calculation of $G(s)$ along the delamination front reduces to a simple post-computation. The extent of the propagation of the delamination area may be established by releasing the node in which the relation (6) is first satisfied, leading to a modification of the delamination front, which in turn requires another equilibrium solution. It follows the fact that the delamination growth analysis must be accomplished iteratively. For simplicity, only the computation of ERR is described here. The study of the propagation for a 3D planar delamination requires the use of nonlinear incremental numerical computation.

The delaminated laminate is represented using two sublaminates (Fig. 2). In this case, the model is called a two-layer plate model. Multilayer plate model in each sublaminate is necessary to achieve sufficient accuracy when the mode components are needed. Sublaminates are modeled using standard shear deformable elements (SHELL181), whereas interface elements can be used for the interface model. Available interface elements (INTER204) are only compatible with solid elements, therefore interface elements are simulated here by coupling CE with spring elements (COMBIN14). Plate and interface models must be described by the same in-plane mesh.

The FE model of the plates adjacent to the delamination plane in proximity of the delamination front is illustrated in Fig. 3. Interface elements model the undelaminated region $\Omega - \Omega_D$ up to the delamination front. The mesh of interface and plate elements must be sufficiently refined in order to capture the high interface stress gradient in the neighborhood of the delamination front, which occurs because high values for interface stiffness must be used to simulate perfect adhesion. The individual ERR at the general node A of the delamination front are calculated using the reactions obtained from spring elements and the relative displacements between the nodes already delaminated and located along the normal direction.

ERRs are computed by using (9), which is a modified version of (8) in order to avoid excessive mesh refining at the delamination front. This leads to the following expressions

$$G_I(A) = \left(\frac{1}{2} \frac{R_A^z \, \Delta w_{B-B'}}{\Delta_n \, \Delta_t} \right), \quad G_{II}(A) = \left(\frac{1}{2} \frac{R_A^n \, \Delta u_{nB-B'}}{\Delta_n \, \Delta_t} \right), \quad G_{III}(A) = \left(\frac{1}{2} \frac{R_A^t \, \Delta u_{tB-B'}}{\Delta_n \, \Delta_t} \right) \quad (9)$$

where R_A^z is the reaction in the spring element connecting node A in the z-direction, $\Delta w_{B-B'}$ is the relative z-displacement between the nodes B and B'. These are located immediately ahead of the delamination front along its normal direction passing through A.

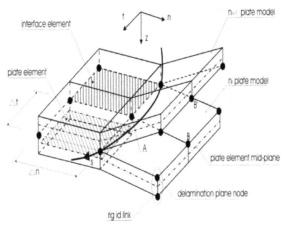

Figure 3. Plate assembly in the neighborhood of the delamination front

Similar definitions apply for reactions and relative displacement related to modes II and III. The characteristic mesh sizes in the normal and tangential directions of the delamination front are denoted by Δ_n and Δ_t. In (9), the same element size is assumed for elements ahead of and behind the delamination front. Value of $\Delta_t /2$ must be used in (9) instead of Δ_t when the node is placed at a free edge.

In order to simplify the FE modeling procedure, it is possible to introduce spring elements only along the delamination front instead of the entire undelaminated region. The perfect adhesion along the remaining portion of the undelaminated region can be imposed by CE. However, when the delamination propagation must be simulated, it is necessary to introduce interface elements in the whole undelaminated region $\Omega - \Omega_D$.

In the next example, the delamination modeling techniques presented so far are applied to analyze typical 3D delamination problems in laminated plates. The ERR distribution along the delamination front are computed for different laminates and loading conditions.

2.2. Finite element modeling and numerical example

One of the most powerful computational methods for structural analysis of composites is the FEM. The starting point would be a "validated" FE model, with a reasonably fine mesh, correct boundary conditions, material properties, etc. (Bathe, 1996). As a minimum requirement, the model is expected to produce stress and strains that have reasonable accuracy to those of the real structure prior to failure initiation. In spite of the great success of the FEM and BEM as effective numerical tools for the solution of boundary-value problems on complex domains, there is still a growing interest in the development of new advanced methods. Many meshless formulations are becoming popular due to their high adaptivity and a low cost to prepare input data for numerical analysis (Guiamatsia et al., 2009).

The results of the delamination analysis of two laminae with different thickness and material are processed in this section. The laminae are fixed on one side and free on the other side. Loads are applied on the free side depending on the analyzed type of delamination.

The upper sublamina has these properties: E_x= 35000 MPa, $E_y = E_z$ = 10500 MPa, G_{yz} = 10500 MPa, $G_{xy} = G_{xz}$ = 1167 MPa, $v_{xy} = v_{xz} = v_{yz}$= 0.3. The lower sublamina has these properties: E_x= 70000 MPa, $E_y = E_z$= 21000 MPa, G_{yz} = 2100 MPa, $G_{xy} = G_{xz}$ = 2333 MPa, $v_{xy} = v_{xz} = v_{yz}$= 0.3. The pair of forces applied on the laminae is T= 1 N/mm and the dimensions of the laminate are: a = 10mm, B = 20 mm , L= 20mm, h_1= 0.5 mm, h_2= 1mm.

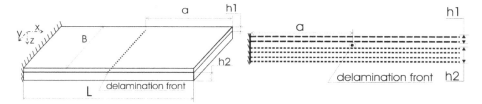

Figure 4. a) Scheme of boundary conditions and laminate dimensions, b) scheme of FEM model

The upper sublaminate is composed by four plates n_u= 4 and the lower by two n_l = 2. The plates are meshed by SHELL181 elements. The zone of mesh refinement has these dimensions 5 * 20 mm, it is centered around of the delamination front which is placed in the middle of the laminate. The interface between the sublaminates is modeled without stiffness for opening displacements and with positive stiffness for closing displacements. The interface between sublaminates is modeled by means of CE (Constrain Equation), since it is easier to apply than beam elements and delamination propagation is not solved. The delamination front is created by spring elements COMBIN 14, in each node of the delamination front by three elements. The stiffness of the spring elements binding the laminae is chosen as k_z a k_{xy} = 10^8 N/mm^3. These elements are oriented in different directions, they were created always from a pair of nodes placed on the surface of the lower sublamina. One of the pair nodes is bounded to the upper plate by means of CE and the second one is bounded to the lower plate. ERR is calculated by using deformation along the delamination front.

Model I

At model I the ERR for delamination type I is analyzed. Model I is loaded with opening forces T of magnitude 1 N/mm, which are parallel to z axis, displayed on Fig. 5a. For the calculation of the ERR the equation (9) was used for the type I. As the biggest ERR is in the middle of the model, it is expected that the beginning of the delamination is in the middle of the model. The distribution of ERR through the width of the laminate is displayed on the Fig.5b.

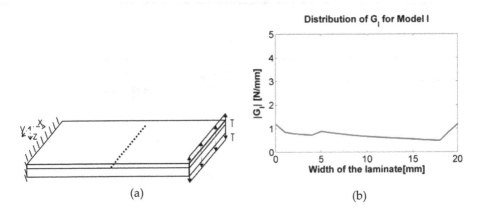

(a)   (b)

Figure 5. Scheme of FEM Model I: a) Forces applied on the laminate model, b) ERR distribution of G_I for delamination type I

Model II

In this model the delamination type II (sliding type)was simulated. The applied forces are parallel to the x axis, Fig. 6 a). Two types of ERR were analyzed in this model, G_{II} in the x direction and G_{III} in the y direction. ERRs were calculated separately for each direction. The reaction of the spring elements are used for the calculation of G_{II} and the y reactions are used for the calculation of G_{III}. Both distributions are displaying the absolute values of ERR, at both distributions the values of ERR are smaller then the values of G_I. The values of G_{II} are in the range of $(0.5, 2) \cdot 10^{-4}$ and the values of $GIII$ are in the range of $(0, 4) \cdot 10^{-5}$ (Fig. 7)

Model III

At this model the delamination type III (tearing type) was analyzed. The geometry of model I was used, with the mesh and its refinement around the delamination front, boundary conditions and the linking between the shell plates, but the direction of the applied forces has changed, Fig.8a). Both types of ERR are analyzed here, G_{II} in the x direction and G_{III} the y direction. The values of ERR are in these ranges: value of GII in the range of $(0, 14) \cdot 10^{-3}$ and the value of $GIII$ in the range of $(0, 0.02)$. It is possible that better results could be achieved by increasing of the number of plate elements layers simulating the sublamina. These models can be also modeled by solid elements, but there is greater number of elements needed for accurately simulating of the stress and ERR gradients. Thereby the number of equations and computing time increase.

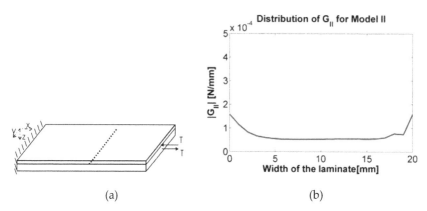

Figure 6. Scheme of FEM Model II: a) direction of loads for delamination type II, b) distribution of ERR for type II delamination of G_{II}

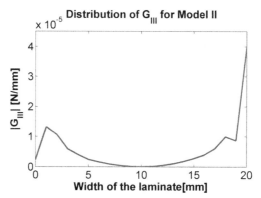

Figure 7. ERR distribution of G_{III} for delamination type II

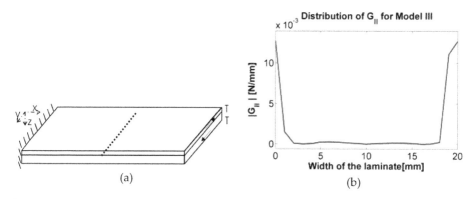

Figure 8. Scheme of FEM Model III: a) definitions of loads for delamination type III, b) distribution of G_{II} for model III

3. Continuum damage mechanics

There are many material modeling strategies to predict damage in laminated composites subjected to static or impulsive loads. Broadly, they can be classified as (Jain & Ghosh, 2009):

- failure criteria approach (Kormaníková, 2011),
- fracture mechanics approach (based on energy release rates),
- plasticity or yield surface approach,
- damage mechanics approach

We consider a volume of material free of damage if no cracks or cavities can be observed at the microscopic scale. The opposite state is the fracture of the volume element. Theory of damage describes the phenomena between the virgin state of material and the macroscopic onset of crack (Jain & Ghosh, 2009; Tumino et al., 2007). The volume element must be of sufficiently large size compared to the inhomogenities of the composite material. In Fig. 9 this volume is depicted. One section of this element is related to its normal and to its area S. Due to the presence of defects, an effective area $\tilde{S}$ for resistance can be found. Total area of defects is therefore:

$$S_D = S - \tilde{S} \tag{10}$$

The local damage related to the direction **n** is defined as:

$$D = \frac{S_D}{S} \tag{11}$$

For isotropic damage, the dependence on the normal n can be neglected, i.e.

$$D = D_n \ \forall n \tag{12}$$

We note that damage D is a scalar assuming values between 0 and 1. For $D = 0$ the material is undamaged, for $0 < D < 1$ the material is damaged, for $D = 1$ complete failure occurs. The quantitative evaluation of damage is not a trivial issue, it must be linked to a variable that is able to characterize the phenomenon.

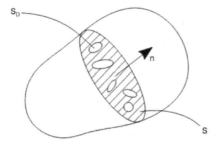

Figure 9. Representative volume element for damage mechanics

We note that several papers can be found in literature where the constitutive equations of the materials are a function of a scalar variable of damage (Barbero, 2008). For the formulation of a general multidimensional damage model it is necessary to generalize the scalar damage variables. It is therefore necessary to define corresponding tensorial damage variables that can be used in general states of deformation and damage (Tumino at al., 2007).

In this part, we focused on presenting the methodology of numerical solving of elastic damage of thin composite plates reinforced by long fibers based on continuum damage mechanics by means of the finite element method.

3.1. Damage model used

The model for fiber-reinforced lamina mentioned next was presented by Barbero and de Vivo (Barbero, 2001) and is suitable for fiber - reinforced composite material with polymer matrix. On the lamina level these composites are considered as ideal homogenous and transversely isotropic. All parameters of this model can be easily identified from available experimental data. It is assumed that damage in principal directions is identical with the principal material directions throughout the damage process. Therefore the evolution of damage is solved in the lamina coordinate system. The model predicts the evolution of damage and its effect on stiffness and subsequent redistribution of stress.

3.1.1. Damage surface and damage potential

Damage surface is similar to the Tsai-Wu damage surface and is defined by tensors **J** and **H** (Barbero, 2008 and it is commonly used for predicting failure of fiber-reinforced lamina with respect to experimental material strength values. Damage surface and damage potential have the form of (Barbero & de Vivo, 2001)

$$g(Y,\gamma) = \sqrt{J_{11}Y_1^2 + J_{22}Y_2^2 + J_{33}Y_3^2} + \sqrt{H_1 Y_1^2 + H_2 Y_2^2 + H_3 Y_3^2} - \left(\gamma + \gamma_0\right) \tag{13}$$

$$f(Y,\gamma) = \sqrt{J_{11}Y_1^2 + J_{22}Y_2^2 + J_{33}Y_3^2} - \left(\gamma + \gamma_0\right) \tag{14}$$

where the thermodynamic forces Y_1, Y_2 and Y_3 can be calculated by means of relations

$$Y_1 = \frac{1}{\Omega_1^2}\left(\frac{\overline{S}_{11}}{\Omega_1^4}\sigma_1^2 + \frac{\overline{S}_{12}}{\Omega_1^2 \Omega_2^2}\sigma_1\sigma_2 + \frac{\overline{S}_{66}}{\Omega_1^2 \Omega_2^2}\sigma_6^2\right)$$

$$Y_2 = \frac{1}{\Omega_2^2}\left(\frac{\overline{S}_{22}}{\Omega_2^4}\sigma_2^2 + \frac{\overline{S}_{12}}{\Omega_1^2\Omega_2^2}\sigma_1\sigma_2 + \frac{\overline{S}_{66}}{\Omega_1^2\Omega_2^2}\sigma_6^2\right) \tag{15}$$

$$Y_3 = 0$$

where stresses and components of matrix $\overline{S}$ are defined in the lamina coordinate system. Matrix $\overline{S}$ gives the strain-stress relations in the effective configuration (Barbero, 2007).

Equation (13) and (14) can be written for different simple stress states: tension and compression in fiber direction, tension in transverse direction, in-plane shear. Tensors J and H can be derived in terms of material strength values.

3.2.1. Hardening parameters

In the present model for damage isotropic hardening is considered and the hardening function was used in the form of

$$\gamma = c_1 \left[\exp\left(\frac{\delta}{c_2}\right) + 1 \right] \tag{16}$$

The hardening parameters γ_0, c_1 and c_2 are determined by approximating the experimental stress-strain curves for in-plane shear loading. If this curve is not available, we can reconstruct it using function

$$\sigma_6 = F_6 \tanh\left(\frac{G_{12}}{F_6}\gamma_6\right) \tag{17}$$

where F_6 is the in-plane shear strength, G_{12} is the in-plane initial elasticity modulus and γ_6 is the in-plane shear strain (in the lamina coordinate system). This function represents experimental data very well.

3.1.3. Critical damage level

The reaching of critical damage level is dependent on stresses in points of lamina. If in a point of lamina only normal stresses in the fiber direction or transverse direction (i.e., normal stress in lamina coordinate system) occur, then simply comparing the values of damaged variables with critical values of damage variables for given material at this point is sufficient. The damage has reached critical level if at least one of the values D_1, D_2 in the point of lamina is greater or equal to its critical value. If in a given point of lamina also shear stress occurs (in lamina coordinate system), it is additionally necessary to compare the value of the product of $(1 - D_1)(1 - D_2)$ with k_s value for given material. If the value of this product is less or equal to k_s value, the damage has reached a critical level.

3.2. Implementation of numerical method

The Newton-Raphson method was used for solving the system of nonlinear equations. Evolution of damage has been solved using return-mapping algorithm described in (Neto, 2008). The input values are strains and strain increments in lamina coordinate system, state variables D_1, D_2, and δ in integration point from the start of last performed iteration, $\bar{C}$ matrix (gives the stress-strain relations in the effective configuration (Barbero, 2007) and damage parameters related to damage model. The output variables are D_1, D_2, and δ, stresses and strains in lamina coordinate system in this integration point at the end of the last performed iteration. Another output is constitutive damage matrix $\mathbf{C}_{ED}$ in lamina coordinate system, which reflects the effect of damage on the behavior of structure. Flowchart of this algorithm is described in Fig. 10.

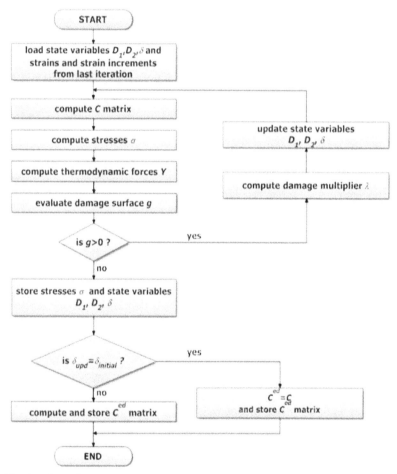

Figure 10. Flowchart of return-mapping algorithm used for solving damage evolution in particular integration points

3.3. Numerical example

One problem for two different materials was simulated in order to study the damage of laminated fiber-reinforced composite plates. The composites are reinforced by carbon fibers embedded in epoxy matrix. The simply supported composite plate with laminate stacking sequence of [0, 45, -45, 90]s with dimensions of 125×125×2.5 mm was loaded by transverse force $F = -4000$ N in the middle of the plate. Own program created in MATLAB language was used for this analysis. Four-node layered plate finite elements based on Kirchhoff (classical) plate theory were used.

Material properties, damage parameters and hardening parameters and critical damage values are given in Table 1 - Table 3. The parameters J_{33} and H_3 are equal to zero. The plate model was divided into 8×8 elements and was analyzed in fifty load substeps. The linear static analysis shows that the largest magnitudes of stress are in parallel direction with fibers and transverse to fibers and they occur in the outer layers in the middle of the plate.

	E_1[GPa]	E_2[GPa]	G_{12}[GPa]	υ_{12}
M30/949	167	8.13	4.41	0.27
M40/948	228	7.99	4.97	0.292

Table 1. Material properties

	J_1	J_2	H_1	H_2	Y_0	c_1	c_2
M30/949	$0.952.10^{-3}$	0.438	$25.585.10^{-3}$	$21.665.10^{-3}$	0.6	0.30	0.395
M40/948	$2.208.10^{-3}$	0.214	$10.503.10^{-3}$	$8.130.10^{-3}$	0.12	0.10	0.395

Table 2. Damage and hardening parameters

	D_{1t}^{cr}	D_{1c}^{cr}	D_{2c}^{cr}	k_s
M30/949	0.105	0.111	0.5	0.944
M40/948	0.105	0.111	0.5	0.908

Table 3. Critical values of damage variables

The largest magnitudes of shear stress occur in the outer layers in the corner nodes. The largest magnitudes of stress in layers 2, 3, 6 and 7 occur in the center of the plate. According to the results of linear static analysis it can be expected that damage will reach the critical level in some of the above points.

Fig. 11 shows the analysis results of elastic damage of the plate made from material M30/949. Fig. 11a shows the evolution of individual stress components in dependence on strains (both in lamina coordinate system) in the midsurface of layer 1 (first layer from the bottom) in integration point (IP) 1 (in element 1, nearest to the corner). Fig. 11b shows the evolution of individual stress components in the midsurface of layer 2 in IP 872 (in element 28, nearest to the center of the plate). Fig. 12 plots described damage variables evolution in these IPs. The analysis results show that reaching the critical level is caused not by normal stresses in the lamina coordinate system, but by shear stress (in the lamina coordinate system). The analysis results of the plate made from material M30/949 show that for given load the critical level of damage was reached in layers 2 and 7 in the center of the plate and its vicinity. In IPs that are closest to the center of the plate in these layers, the critical level of damage was reached between 13th and 14th load substep. However, it is not postulated that used damage model predicts failure: it only predicts damage evolution and its effect on stiffness and consequent stress redistribution (Barbero & de Vivo, 2001). In some cases, failure can occur before the critical level of damage is reached. For plate made from material M30/949 load F=-1096 N is already critical.

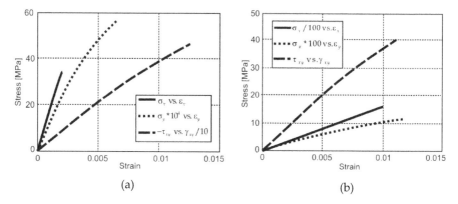

Figure 11. Stress and strain evolution for plate from material M30/949 in (a) IP 1, (b) IP 872

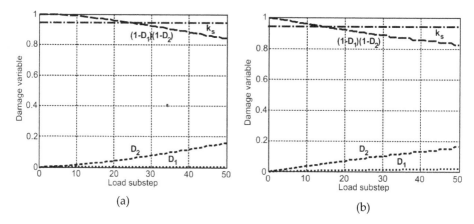

Figure 12. Damage variables evolution for plate from material M30/949 in (a) IP 1, (b) IP 872

Fig. 13 shows the analysis results of elastic damage of the plate made from material M40/948. Fig. 13a shows the evolution of individual stress components in dependence on strains (both in lamina coordinate system) in the midsurface of layer 1 in IP 868 (in element 28, nearest to the center of the plate). Fig. 13b shows the evolution of individual stress components in the midsurface of layer 2 in IP 872 (in element 28, nearest to the center of the plate). The results show that reaching the critical level of damage will be also caused by shear stress in lamina coordinate system. However, the critical damage level was reached in layers 1 and 7 in the center of plate and its vicinity. The critical level of damage was reached between 12th and 13th load substep in the nearest IPs. The critical level of damage would be also reached in the center of the plate and its vicinity in layers 2 and 7 (in the nearest IPs it would be reached between 16th and 17th load substep) and also in layers 3 and 6 (in the nearest IPs it would be reached between 27th and 28th load substep). Fig. 14 shows damage variables evolution in IP 868 and IP 872. For plate made from material M40/948 load F = - 990 N is already critical.

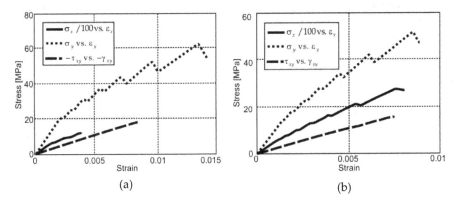

Figure 13. Stress and strain evolution for plate from material M40/948 in (a) IP 868, (b) IP 872

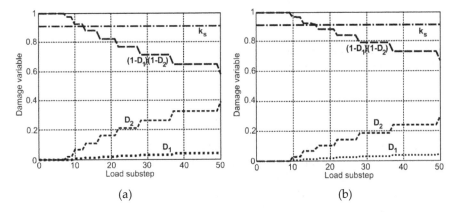

Figure 14. Damage variables evolution for plate from material M40/948 in (a) IP 868, (b) IP 872

4. Conclusion

The methodology of delamination calculation in laminated plates was applied in this chapter. The analyses shows that if mixed mode conditions are involved, a double plate model is suitable to accurately capture the mode decomposition in region near the midpoint of delamination front. The solution converges quickly because a small number of plates is needed to obtain a reasonable approximation

Damage model presented in this chapter has been utilized in this solution. This damage model is suitable for elastic damage of fiber-reinforced composite materials with polymer matrix. The postulated damage surface reduces to the Tsai-Wu surface in stress space. The problem of elastic damage is considered as material nonlinearity, so we get system of nonlinear equations. The Newton-Raphson method has been used for solving this system of nonlinear equations. Evolution of damage has been solved using the return-mapping algorithm. Flowchart of this algorithm was also presented. Numerical example of one problem for two different materials was presented next. Own program created in MATLAB language was used for this analysis. Four-node layered plate finite elements based on Kirchhoff (classical) plate theory were used. The analysis results show that change of material as well as the presence and values of shear stress have significant influence on the evolution of damage as well as on location of critical damage and load at which the critical level of damage will be reached. Critical damage level has not necessary to be reached in places with maximum magnitude of equivalent stress, but can be reached in other places.

Author details

Milan Žmindák and Martin Dudinský
University of Žilina, Slovakia

Acknowledgement

The authors gratefully acknowledge the support by the Slovak Grant Agency VEGA 1/1226/12 and Slovak Science and Technology Assistance Agency registered under number APVV-0169-07.

5. References

Ansys v.11, *Theory manual* (2007), ANSYS, Inc., Southpointe, PA.

Azouaoui, K. et al. 2010. Evaluation of impact fatigue damage in glass/epoxy composite laminate. *International Journal of Fatigue*, Vol. 32, No. 2, pp. 443-452, ISSN 0142-1123

Bathe, K. J. (1996). *Finite Element Procedures*, Prentice Hall, ISBN 0-13-301458-4, New Jersey

Barbero, E. J. (2007). *Finite Element Analysis of Composite Materials*, CRC Press, ISBN 1-4200-5433-3, Boca Raton

Barbero, E. J. & De Vivo, L. (2001). A Constitutive Model for Elastic Damage in Fiber-Reinforced PMC Laminae. *International Journal of Damage Mechanics*. Vol. 10, No. 1, pp. 73-93, ISSN 1056-7895

Carrera, E. (2002). Theories and Finite Elements for Multilayered, Anisotropic, Composite Plates and Shells. *Arch. Comput. Meth. Engng.* Vol. 9, Nr. 2, 2002, pp. 87-140, ISSN 1134-3060.

Clegg, R. A. et al. (2006). Hypervelocity impact damage prediction in composites: Part I-material model and characterisation. *International Journal of Impact Engineering*. Vol. 33, No. 1-12, pp. 190-200, ISSN 0734-743X

Cui W, Wisnom M. A. (1993). Combined stress-based and fracture mechanics-based model for predicting delamination in composites. *Composites* 1993; Vol. 24, pp. 467–74, ISSN 1359-835X.

Dávila, C.G., Camanho, P.P., Rose, Ch. A. (2005), Failure criteria for FRP Laminates. *Journal of Composite Materials*. Vol. 39, No.4, pp. 323-345. ISSN 0021-9983

Elmarakbi, A., Hu, N., Fukunaga, H., (2009), Finite element simulation of delamination growth in composite materials using LS-DYNA. *Composites Science & Technology*, 69 pp. 2383-2391, ISSN 0266-3538

Fenske, M.T., Vizzini, A.J. (2001), The inclusion of in-plane stresses in delamination Criteria. *Journal of Composite Materials*. Vol. 35, No.15, pp. 1325-1342. ISSN 0021-9983

Gayathri, P. et al. (2010). Effect of matrix cracking and material uncertainty on composite plates. *Reliability Engineering & System Safety*. Vol. 95, No. 7, pp. 716-728, ISSN 0951-8320

Guiamatsia, I. et al. (2009). Element-free Galerkin modelling of composite damage. *Composites Science and Technology*. Vol. 69, No. 15-16, pp. 2640-2648, ISSN 0266-3538

Chung Deborah, D.L. (2003). *Composite Materials: Functional Materials for Modern Technology*, ISBN 1-85233-665-X, Springer.

Iannucci, L. & Ankersen, J. (2006). An energy based damage model for thin laminated composites. *Composites Science and Technology*, Vol. 66, No. 7-8, pp. 934-951. ISSN 0266-3538

Jain, J.R. & Ghosh, S. (2009). Damage Evolution in Composites with a Homogenization-based Continuum Damage Mechanics Model. *International Journal of Damage Mechanics*, Vol. 18, No. 6, pp. 533-568, ISSN 1056-7895

Kormaníková, E., Žmindák, M., Riecky, D. (2011). Strength of composites with fibres. In: *Computational Modelling and Advanced Simulations*, Murín, J., et al., pp. 167-184, ISBN 978-94-007-0316-2 Springer Science + Business Media B.V., Dordercht

Laš, V. & Zemčík, R. (2008). Progressive Damage of Unidirectional Composite Panels. *Journal of Composite Materials*, Vol. 42, No. 1, pp. 25-44, ISSN 0021-9983

Mishnaevsky, L., Jr (2007). Computational mesomechanics of Composites, Numerical analysis of the effect of microstructures of composites on their strength and damage resistance.

Neto, E.A. de Souza, Peric, D., Owen, D.R.J. (2008). *Computational Methods for Plasticity: Theory and Applications*, ISBN 978-0-470-69452-7, John Wiley & Sons, Ltd.

Reddy, J.N., Miravete, A. (1995), Practical analysis of composite laminates, ISBN 0-8493-9401-5, CRC Press

Riccio, A. & Pietropaoli, E. (2008). Modeling Damage Propagation in Composite Plates with Embedded Delamination under Compressive Load. *Journal of Composite Materials*, Vol. 42, No. 13, pp. 1309-1335, ISSN 0021-9983

Sládek, J., Sládek, V., Jakubovičová, L (2002). *Application of Boundary Element Methods in Fracture Mechanics*, ISBN 80-968823-0-9, University of Žilina, Faculty of Mechanical Engineering, Žilina

Shu, X. (2006). A generalised model of laminated composite plates with interfacial damage. *Composite Structures*, Vol. 74, No. 2, pp. 237-246, ISSN 0263-8223

Tiberkak, R. et al. (2008). Damage prediction in composite plates subjected to low velocity impact. *Composite Structures*, Vol. 83, No. 1, pp. 73-82, ISSN 0263-8223

Tumino, D., Campello, F., Catalanotti, G. (2007). A continuum damage model to simulate failure in composite plates under uniaxial compression. *Express Polymer Letters* 1, No.1, pp. 15-23, ISSN 1788-618X

Zhang, Z., Wang, S., (2009). Buckling, post-buckling and delamination propagation in debonded composite laminates. Part 1: Theoretical development, *Composite Structures*, Vol. 88, pp. 121-130, ISSN: 0263-8223.

Žmindák, M., Riecky, D., Danišovič, S. (2010). Finite element implementation of failure and damage models for composite structures, *Proceedings of the XV international conference " Machine Modelling and Simulation*, Warszaw university of technology, Krasiczyn, August, 2010.

Žmindák, M., Riecky,D., Soukup, J. (2010). Failure of Composites with Short Fibers. Communications. 4/2010, pp. 33-39, ISSN-13354205.

Chapter 7

Modelling of Fracture of Anisotropic Composite Materials Under Dynamic Loads

Andrey Radchenko and Pavel Radchenko

Additional information is available at the end of the chapter

1. Introduction

At present, the wide application of the materials with preset directivity of properties in various fields of engineering defines the increased interest to the investigations of anisotropic materials behaviour under various conditions. But in Russia as well as abroad such investigations are conducted mainly for static conditions. The behaviour of anisotropic materials under dynamic loads is practically not investigated. This is especially the case with experimental investigations as well as with mathematical and numerical modelling. The impact interaction of the solids in a wide range of kinematic and geometric conditions is the complex problem of mechanics. The difficulties, connected with the theoretic study of the fracture and deformation of materials on impact by analytical methods force to introduce member of simplifying hypothesis which distort the real picture in majority of cases. In this connection it should be accepted that the leading role in the investigation of phenomena, connected with high-speed interaction of solids belongs to the experimental and numeric investigations at present. The investigations of the material damage under impact show that the fracture mechanisms change with the interaction conditions. The experiments strongly testifiers that in a number of case the resulting fracture is determined by the combination of several mechanisms. But in the experiments we fail to trace sequence, operation time and the contribution of various fracture mechanisms. Besides, the distractions, obtained at the initial stages of the process can't always be identified in the analysis of the resulting fracture of the materials. For anisotropic materials the strength itself is multivalued and uncertain notion due to the polymorphism of behaviour of these materials under the load. The limiting state of anisotropic bodies may be of different physical nature in dependence on load orientation, stressed state type and other factors. The dependence of the physical nature of limiting states is revealed in the sturdy of the experimental data. The investigations of hydrostatic pressure effect upon the strength of isotropic materials show that comprehensively the compression exerts a weak action on the resistance of isotropic materials under static loads. Therefore, the classic theories of strength, plasticity and creep are based on assumption about the lack of the effect of fall stress tensor upon strength isotropic materials. In the experiments with anisotropic materials it was state that flow phenomenon may arise only under the action of hydrostatic pressure. Upon the materials, strength is due to the anisotropy. The shape of anisotropic bodies changes under the action of hydrostatic pressure. If these changes reach such values that, they don't disappear under relief, the limiting state should come. Therefore, the postulate of classic strength that hydrostatic pressure can't transfer the material to the dangerous state is not valid for anisotropic materials. The phenomenological approach to the investigation of the dynamics of deformation and fracture of anisotropic and isotropic materials is used in the project. The phenomenological approach to the materials strength requires that the conditions of the transition into the limiting state of various physical natures should be determinated by one equation (criterion). The necessity of such an approach results from fracture polymorphism, being deduced experimentally. For anisotropic bodies the phenomenological approach has many advantages, since there appears the possibility to use general condition of strength for the material different in composition and technology but similar in symmetry of properties, and also for the materials with substantial anisotropy, for which one and the same stressed state can result in limiting conditions, different in physical nature.

2. Equations of the model

2.1. Basic equations

The system of the equations describing non-stationary adiabatic movements of the compressed media in the Cartesian coordinate system XYZ, includes following equations (Johnson, 1977):

- continuity equation

$$\dot{\rho} + \operatorname{div} \rho \vec{\upsilon} = 0 \tag{1}$$

- motion equation

$$\rho\dot{u} = \sigma_{xx,x} + \sigma_{xy,y} + \sigma_{xz,z}$$
$$\rho\dot{\upsilon} = \sigma_{yx,x} + \sigma_{yy,y} + \sigma_{yz,z} \tag{2}$$
$$\rho\dot{w} = \sigma_{zx,x} + \sigma_{zy,y} + \sigma_{zz,z}$$

- energy equation

$$\dot{E} = \frac{1}{\rho}\sigma_{ij}e_{ij}; \quad i,j = x,y,z \tag{3}$$

Here ρ – density of media; $\vec{\upsilon}$ – velocity vector, u, υ, w – components of velocity vector on axes x, y, z accordingly; σ_{ij} – components of a symmetric stress tensor; E – specific internal energy; e_{ij} – components of a symmetric strain rate tensor; the point over a symbol means a time derivative; a comma after a symbol – a derivative on corresponding coordinate.

The behavior of the aluminum isotropic cylinder at high-velocity impact is described by elastic-plastic media, in which communication between components of strain velocity tensor and components of stress deviator are defined by Prandtl-Reuss equation:

$$2G\left(e_{ij} - \frac{1}{3}e_{kk}\delta_{ij}\right) = \frac{DS^{ij}}{Dt} + \lambda S^{ij}, \ (\lambda \geq 0); \quad \frac{DS^{ij}}{Dt} = \frac{dS^{ij}}{dt} - S^{ik}\omega_{jk} - S^{jk}\omega_{ik} \tag{4}$$

where $\omega_{ij} = \frac{1}{2}\left(\nabla_i \upsilon_j - \nabla_j \upsilon_i\right)$, G – shear modulus. Parameter $\lambda = 0$ at elastic deformation, and at elastic ($\lambda > 0$) is defined by means of a Mises condition:

$$S^{ij}S_{ij} = \frac{2}{3}\sigma_d^2 \tag{5}$$

where σ_d – dynamic yield point. The ball part of stress tensor (pressure) is calculated on the Mi-Gruneisen equation as function of specific internal energy E and density ρ:

$$P = \sum_{n=1}^{3} K_n \left(\frac{V_0}{V} - 1\right)^n \left[\frac{1 - K_0\left(\frac{V_0}{V} - 1\right)}{2}\right] + K_0\rho E \tag{6}$$

where K_0, K_1, K_2, K_3 – constants of material.

2.2. Model of deformation and fracture of anisotropic materials

The behavior of an anisotropic material of targets is described within the limits of elastic-fragile model. Before fracture components of a stress tensor in a target material were defined from equations of the generalized Hooke's law which have been written down in terms of strain rate:

$$\dot{\sigma}_{ij} = C_{ijkl}e_{kl} \tag{7}$$

where C_{ijkl} – elastic constants.

Thus components of a tensor of elastic constants possess, owing to symmetry of stress tensors and strain tensors and presence of the elastic potential, following properties of symmetry:

$$C_{ijkl} = C_{jikl} = C_{ijlk} = C_{ijlk}; \ C_{ijkl} = C_{klij} \tag{8}$$

At transition to another, also orthogonal, coordinate system, elastic constants will be transformed by equations:

$$C'_{abcd} = C_{ijkl} q_{ia} q_{jb} q_{kc} q_{ld} \tag{9}$$

where q_{ij} – cosine of the angle between corresponding axes i and j. In three-dimensional space transformation of the component of a tensor of the fourth rank demands summation of the compositions, containing as multipliers 4 cosines of angles of rotation of axes.

Fracture of an anisotropic material is described within the limits of model with use of Tsai-Wu fracture criterion with various ultimate strengths of pressure and tension (Tsai & Wu, 1971). This criterion, which has been written down by scalar functions from components of a stress tensor, has the following appearance:

$$f(\sigma_{ij}) = F_{ij}\sigma_{ij} + F_{ijkl}\sigma_{ij}\sigma_{kl} + \ldots \geq 1; \quad i, j, k, l = 1, 2, 3 \tag{10}$$

Here F_{ij} and F_{ijkl} are components of tensor of the second and the fourth rank respectively, and obey transformation laws:

$$F'_{ab} = F_{ij} q_{ia} q_{jb}; \quad F'_{abcd} = F_{ijkl} q_{ia} q_{jb} q_{kc} q_{ld} \tag{11}$$

Components of tensors of strength for criterion are defined by following equations:

$$F_{ii} = \frac{1}{X_{ii}} - \frac{1}{X'_{ii}}; \quad F_{iiii} = \frac{1}{X_{ii} X'_{ii}}; \quad F_{ij} = \frac{1}{2}\left(\frac{1}{X_{ij}} - \frac{1}{X'_{ij}}\right); \quad F_{ijij} = \frac{1}{4 X_{ij} X'_{ij}}; \quad i \neq j \tag{12}$$

where X_{ii}, X'_{ii} – limits of strength on pressure and tension along the direction i; X_{ij}, X'_{ij} – shear strength along the two opposite directions with $i \neq j$. Coefficients F_{1122}, F_{2233}, F_{3311} are defined at carrying out the experiments on biaxial tension in planes 1–2, 2–3, 1–3 accordingly. The remained coefficients are defined similarly at combined stressing in corresponding planes (Radchenko et al., 2012).

It is supposed that fracture of anisotropic materials in the conditions of intensive dynamic loads occurs as follows:

- if strength criterion is violated in the conditions of pressure $(e_{kk} \leq 0)$, the material loses anisotropy of properties, and its behaviour is described by hydrodynamic model, thus the material keeps its strength only on pressure; the stress tensor becomes in this case spherical ($\sigma_{ij} = -P$);
- if the criterion is violated in the conditions of tension $(e_{kk} > 0)$, the material is considered completely fractured, and components of a stress tensor are appropriate to be equal to zero ($\sigma_{ij} = 0$).

Pressure in orthotropic materials of targets is calculated by means of the equation of a condition (Kanel et al., 1996):

$$P = \left[\exp\left(4\beta\frac{V_0 - V}{V_0}\right) - 1\right]\frac{\rho_0 \alpha^2}{4\beta} \tag{13}$$

Here ρ_0 is initial density of a material; V_0, V – relative initial and current volumes. Coefficients of the given equation are calculated from a shock adiabat: $D = \alpha + \beta u$, where α =1400 m/s, β =2.25, and u – mass velocity.

2.3. Initial and boundary conditions

It is considered (fig. 1) a three-dimensional task of high-speed interaction of compact (diameter of the projectile is equal to its height) cylindrical projectile (area D_1) with one or several targets (areas D_2, D_3, D_4). In this paper we consider the materials with the following mechanical characteristics (Ashkenazi & Ganov, 1980): Steel St3 with ρ_0 =7850 kg/m³, E =204 GPa, μ =0.3, G =79 GPa, $\sigma_{0.2}$ =1.01 GPa, K_0 =1.91, K_1 =153 GPa, K_2 =176 GPa, K_3 =53.1 GPa; aluminum with ρ_0 =2700 kg/m³, E =70 GPa, μ =0.3, G =27 GPa, $\sigma_{0.2}$ =310 MPa, K_0 =2.13, K_1 =74.4 GPa, K_2 =53.2 GPa, K_3 =30.5 GPa; organoplastic with ρ_0 =1350 kg/m³, E_1 =48.6 GPa, E_2 =21.3 GPa, E_3 =7.1 GPa, μ_{12} =0.28, μ_{23} =0.26, μ_{31} =0.037, G_{12} =930 GPa, G_{23} =900 GPa, G_{31} =850 GPa, X_{11} =2.67 GPa, X_{22} =1.18 GPa, X_{33} =0.395 GPa, X'_{11} =0.37 GPa, X'_{22} =0.5 GPa, X'_{33} =1.94 GPa, X_{12} =0.975 GPa, X_{23} =0.8 GPa, X_{31} =0.607 GPa, $X_{11}^{(12)}$ =2.3 GPa, $X_{11}^{(31)}$ =2 GPa, $X_{22}^{(12)}$ =1 GPa, $X_{22}^{(23)}$ =0.9 GPa, $X_{33}^{(23)}$ =0.35 GPa, $X_{33}^{(31)}$ =0.31 GPa, c_x =6000 m/s, c_y =3970 m/s, c_z =2300 m/s. The meeting corner (between a normal to a target and a longitudinal axis of the projectile) made a corner $\alpha = 0°$ (normal impact).

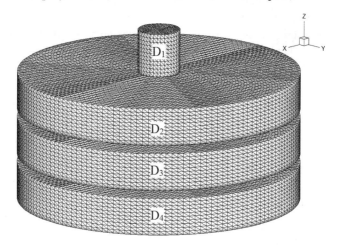

Figure 1. Three-dimensional formulation of the task

Initial conditions ($t = 0$):

$$\sigma_{ij} = E = u = v = 0, \quad w = v_0, \quad i, j = x, y, z, \quad x, y, z \in D_1 \tag{14}$$

$$\sigma_{ij} = E = u = v = w = 0, \quad i, j = x, y, z, \quad x, y, z \in D_2, D_3, D_4 \tag{15}$$

$$\rho = \rho_i, \quad x, y, z \in D_i, \quad i = 1, 2, 3, 4 \tag{16}$$

Boundary conditions:

On free surfaces conditions of free border are realized:

$$\overline{T}_{nn} = \overline{T}_{n\tau 1} = \overline{T}_{n\tau 2} = 0 \tag{17}$$

On contact surface sliding condition without a friction is realized:

$$\overline{T}_{nn}^{+} = \overline{T}_{nn}^{-}, \ \ \overline{T}_{n\tau}^{+} = \overline{T}_{n\tau}^{-} = \overline{T}_{ns}^{+} = \overline{T}_{ns}^{-} = 0, \ \ \overline{\upsilon}_{n}^{+} = \overline{\upsilon}_{n}^{-} \tag{18}$$

Here $\overline{n}$ – a unit vector of a normal to a surface in a considered point, $\overline{\tau}$ and $\overline{s}$ – unit vectors, tangents to a surface in this point, $\overline{T}_{n}$ – a force vector on a platform with a normal $\overline{n}$, $\overline{\upsilon}$ – a velocity vector. The subscripts at vectors $\overline{T}_{n}$ and $\overline{\upsilon}$ also mean projections on corresponding basis vectors; the badge plus "+" characterizes value of parameters in a material on the top border of a contact surface, a badge a minus "–" – on bottom.

The problem solves numerically using the finite element method in the explicit formulation by Johnson G.R. (Johnson, 1977).

3. Check of adequacy of the model

The series of test calculations for check convergence of the solution (independence of the solution from a spatial interval) has been passed. Dependence of the σ_{xx} component in the central point of a target from total number of elements N_E in a calculating grid was considered. The received curve shows fast convergence of the decision at a grid compaction (fig. 2).

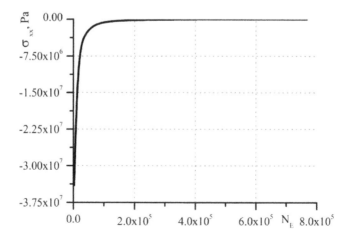

Figure 2. Convergence of the solution

3.1. Experiments with isotropic materials

In connection with a considerable quantity of works on a problem of impact on rigid target, numerical experiments for check of the numerical methods have been made. The problem about normal impact of the cylindrical projectile with length L_0 =23.47 mm and diameter D_0 =7.62 mm on rigid target with initial velocity v_0 is considered. Material of the projectile – steel of St3 mark. In the table 1 the received results of calculations in comparison with experiment and the data of calculations from M.L. Wilkins works on residual length of the drummer L are presented at various velocities of impact (Wilkins & Giunan, 1973). In the table 1 δ_l is the relative divergence between spent calculations and experiments.

v_0, m/sec	Experiment, $\dfrac{L}{L_0}$	M.L. Wilkins, $\dfrac{L}{L_0}$	Calculation, $\dfrac{L}{L_0}$	δ_l, %
175	0.911	0.911	0.915	0.4
252	0.842	0.842	0.839	0.4
311	0.766	0.766	0.760	0.8
402	0.635	0.667	0.619	2.5

Table 1. Comparison with experiments and calculations

For check of adequacy of model the number of comparisons of numerical calculations with experimental data has been spent. Table 2 shown the results of experiments and calculations at the interaction of the steel projectile with mass of 20 grams with the glass-fibre plastic isotropic targets (ρ_0 =1930kg/m³). The following symbols are introduced into the table: h is the target thickness, v_0 is the initial projectile velocity, v_1 is the post-perforation velocity of projectile, ε is the relative decrease of the projectile height after target piercing, δ_v is the relative divergence between post-perforation velocity of projectile in the experiment and calculation.

h, mm	v_0, m/sec	Experiment v_1, m/sec	Calculation v_1, m/sec	δ_v, %
5	992	928	970	6.3
9	1163	1013	950	6.2
14	1064	812	780	4.0

Table 2. Comparison with experiments for isotropic targets

3.2. Experiments with anisotropic materials

A number of experiments of penetration for check of the offered model of behavior of anisotropic materials was similarly carried out (Radchenko et al., 1999). The results on beyond barrier speeds of striker at the interaction of 20-gram striker with transtropic barrier are presented in table 3.

h, mm	v_0, m/s	Experiment v_1, m/s	Calculation v_1, m/s	δ_v, %
26	1054	698	640	8.3
26	1077	695	638	8.2
18	1012	897	836	6.8
18	956	838	792	5.5

Table 3. Comparison with experiments for anisotropic targets

Comparison of numerical and experimental results allows conclude that the proposed model satisfactorily describes the process of breaking through the isotropic and transtropic plates. A deviation of calculated values of the post-penetration velocities from the experimental values does not exceed 8.5%.

4. Deformation and fracture of anisotropic composite materials and designs under dynamic loading

4.1. Pulse effect

Problems of dynamic deformation of the ball from organoplastic under effect of omnidirectional compression pulse are considered in three–dimensional statement. Homogeneous orthotropic ball with diameter of 10 mm was subjected to compression with pulse pressure of 1 GPa during 3 μsec (fig. 3). It is supposed that the failure of anisotropic material in condition of intense dynamic loads happens in accordance with (Radchenko et al., 1999). The material of the ball is orthotropic organoplastic. Already at the time moment of 0.6 μsec in ZX cross section (where there is most significant difference between characteristics), the distribution of stress and field of velocities (fig. 4) illustrate the origin of heterogeneous picture of Strain-stress State of the ball. In ZY section the distribution of stress as well as field of velocities is close to one–dimension Strain-stress State of isotropic ball under the effect of omnidirectional compression (fig. 5). At this moment, the stresses achieves maximum values (-2 GPa) near at ball poles on Z axis.

In this case the ball failure arises in the region of maximum stresses. With time the Strain-stress State of anisotropic ball differs from Strain-stress State of isotropic ball more strongly. Isolines of stress and field of rates at the shown in fig. 6, fig. 7. Evident sliding lines are seen in fig. 6a and fig. 7a. Only σ_x distribution in ZY cross section (fig. 7b) is closest to the distribution of stress in isotropic ball, maximum stress are achieved in the centers.

Up to 1 μsec in all directions, mass velocities are directed into the insight of the ball, but with of 1 μsec, in direction X rates change the sign and the increase in ball size begins in this direction. In other directions, mass velocities are directed inside of the ball up to the moment of load removal (3 μsec). This, maximum decrease in ball dimensions in X directions is 11% and is achieved to 1 μsec, and in Z direction it is 24% and is achieved in 3 μsec.

Substantial change in the shape of the ball is observed to the moment of cessation of compression pulse action (fig. 8) it acquires dump – bells shape due to the compression along Z axis.

The expansion of the ball in all directions begins after the release of the load. Fig. 9 shows the field of velocities in 6 µsec, to that moment the material of the ball has been completely fractured.

The ball under the effect of omnidirectional compression pulse may transform not to ellipsoid but to the dump – bells under the certain relations of its mechanical characteristics of the value of pressure pulse.

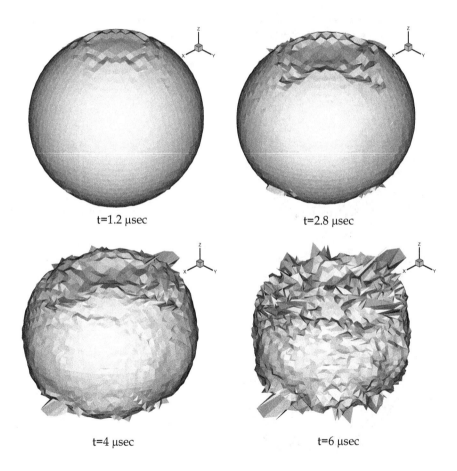

t=1.2 µsec   t=2.8 µsec

t=4 µsec   t=6 µsec

Figure 3. Volume configurations of orthotropic ball at loading by the pulse of pressure: $P = P_0$ if $t \leq \tau$ and $P = 0$ if $t > \tau$. $P = 1$ GPa, $\tau = 3$ µsec

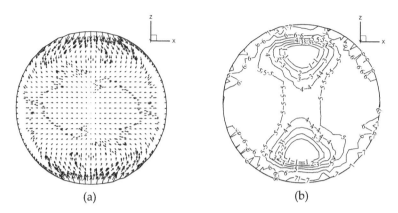

Figure 4. Field of mass velocities (a) and distribution of isolines of stress σ_x (b). t = 0.6 µsec.1: -2, 2: -1.8, 3: -1.6, 4: -1.4, 5: -1.2, 6: -1, 7: -0.8, 8: -0.6 GPa

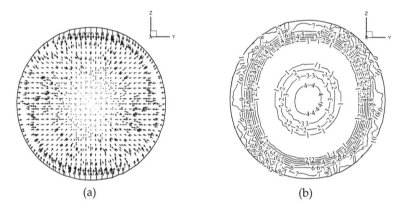

Figure 5. Field of mass velocities (a) and distribution of isolines of stress σ_x (b). t = 0.6 µsec. 1: -2, 2: -1.8, 3: -1.6, 4: -1.4, 5: -1.2, 6: -1, 7: -0.8, 8: -0.6 GPa

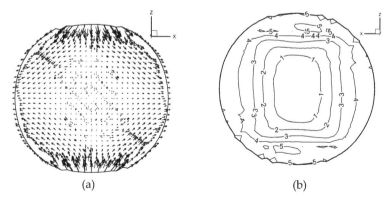

Figure 6. Field of mass velocities (a) and distribution of isolines of stress σ_x (b). t = 1.2 µsec. 1: -2.4, 2: -2, 3: –1.6, 4: –1.2, 5: –0.8, 6: –0.4, 7: 0, 8: 0.4, 9: 0.8, 10: 1.2 Gpa

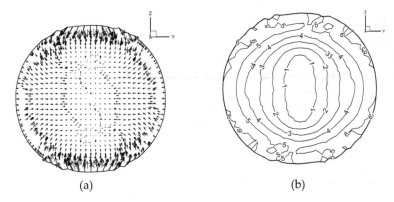

| (a) | (b) |

Figure 7. Field of mass velocities (a) and distribution of isolines of stress σ_x (b). $t = 1.2$ μsec. 1: -2.4, 2: -2, 3: –1.6, 4: –1.2, 5: –0.8, 6: –0.4, 7: 0, 8: 0.4, 9: 0.8, 10: 1.2 GPa

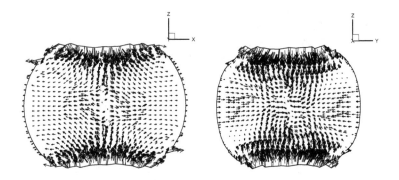

Figure 8. Field of mass velocities. $t = 3$ μsec.

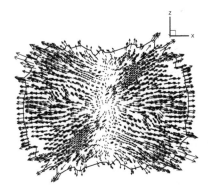

Figure 9. Field of mass velocities. $t = 6$μsec.

4.2. Influence of orientation of elastic and strength properties on fracture of anisotropic materials under dynamic loading

We studied the penetration of a barrier with the initial orientation of the properties, as well as penetration of a barrier with properties reoriented by 90° about the axis OY. In the direction of the axis Z initial material has the highest strength on compression and the lowest strength on tension. Refocused the material, on the contrary, along the Z axis has the lowest compressive strength and the highest tensile strength. In addition to the various limits of the tensile and compression on the dynamics of fracture will affect significantly the velocity of propagation of waves of compression and unloading, which in an anisotropic material depend on the direction.

A material of the projectile is isotropic steel, a material of targets – orthotropic organoplastic. Orientation of properties of orthotropic material changes by turn of axes of symmetry of an initial material round an axis on an angle $\beta = 90°$.

On fig. 10 and fig. 11 configurations of the projectile and targets with distribution of isolines of relative volume of fractures for various velocities of interaction at the moment of time t =40 μsec are presented. To the left of a symmetry axis configurations for initial orientation of a material of a target, to the right – for the reoriented material are given.

For a case of initial orientation of properties of organoplastic at velocity 50 m/s (fig. 10a, to the left of a symmetry axis) on an obverse surface of a target on perimeter of the projectile and on a contact surface in the target center the conic zones of fracture focused at an angle 45° to a direction of impact are formed. These zones arise in an initial stage of interaction at the expense of action of tensile stress in the unloading waves extending from an obverse surface of a target and a lateral surface of the projectile. The further development of these zones of fracture is caused by action of tensile stress as a result of introduction of the projectile. At initial velocity 50 m/s there is no perforation of a target. To t =30 μsec velocity of the projectile reduces to zero and the kickback of the projectile from a target is observed. Values of vertical component of the velocity of the center of projectile weights υ_z and a part of the fractured material of a target are presented in table 4 (at tension D_t and pressure D_p at the moment of time t =50 μsec).

υ_0 , m/s	50		100		200		400	
β	0°	90°	0°	90°	0°	90°	0°	90°
υ_z ,m/s	−5.17	4.08	9.05	12.97	49.97	127.6	191.55	303.69
D_t	0.012	0.005	0.056	0.011	0.162	0.128	0.502	0.282
D_p	0.006	0.002	0.043	0.029	0.104	0.021	0.112	0.019

Table 4. Velocity of the center of projectile weights and part of the break material in targets

In case of the reoriented material (fig. 10a, to the right of a symmetry axis) a picture of development of fracture is qualitative other. In this case, strength of a material on pressure in a direction of axis Z (an impact direction) is minimal. It leads to that the material break in the wave of pressure formed at the moment of impact and extending on a thickness of a target. Penetration of the projectile thus occurs in already weakened material. Though perforation in this case also isn't present, the projectile gets deeply, and its full braking is observed in 50 μsec. With increase in velocity of impact the volume of areas of fracture grows. At velocity 100 m/s (fig. 10b) fracture areas extend to a greater depth on a thickness of a target. And for an initial material of a target the marked orientation (45°) was kept only by a crack extending from an obverse surface on perimeter of the projectile. The crack located near to an axis of symmetry isn't identified any more. It is caused by that with increase in velocity of impact the amplitude of the pressure wave grows – its size is already sufficient for material fracture in the top half of target.

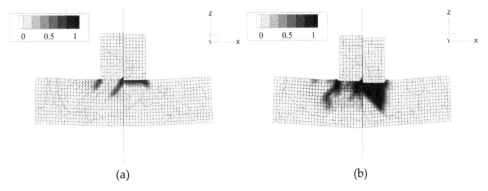

(a)   (b)

Figure 10. Relative volume of fractures in target. υ_0 =50 m/s (a) and υ_0 =100 m/s (b), t=40 μsec.

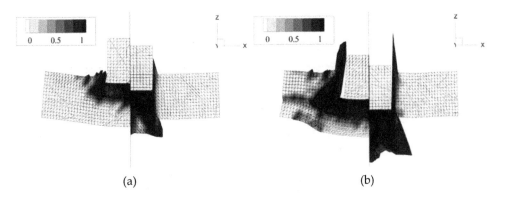

(a)   (b)

Figure 11. Relative volume of fractures in target. υ_0 =200 m/s (a) and υ_0 =400 m/s (b), t=40 μsec.

In case of the reoriented material the unloading wave extending from a back surface of a barrier, lowers level of compressive stresses that leads to smaller distribution of fracture area on a thickness near to a symmetry axis (fig. 10b). For velocity 100 m/s also it is not observed perforation of targets, thus in case of an initial material velocity of the projectile reduced to zero at 45 μsec, in case of the reoriented material – at 60 μsec.

For velocities of impact 200 m/s and above (fig. 11) it is already observed perforation of targets from both types of materials. But thus the plate from an initial material greater maintains resistance to penetration of the projectile in comparison with a plate from the reoriented material. For example, at initial velocities 200 m/s (fig. 11a) and 400 m/s (fig. 11b) post-perforation velocity of the projectile after perforation of plates from an initial material makes 37 m/s and 187 m/s accordingly, and post-perforation velocity after perforation plates of the reoriented material 125 m/s and 300 m/s. Greater resistance to penetration of the projectile in plates from an initial material is caused by a various picture of fracture which is defined by orientation of elastic and strength properties in relation to external loading. For velocities of impact above 200 m/s there is fracture of the reoriented material in the unloading wave extending from a back surface of a target (fig. 11), that increases volume of the break material in front of the projectile, essentially reducing resistance to penetration. Such dynamics of fracture become clear by various velocities of wave distribution in the initial and reoriented materials.

In an initial material velocity of wave distribution the greatest in a direction of an axis X – perpendicular to an impact direction, therefore unloading waves from an obverse surface of a target and a lateral surface of the projectile lower stresses in a pressure wave to its exit on a back surface that doesn't lead to material fracture in a pressure wave in the bottom half of plate and a unloading wave from a back surface of the barrier having small amplitude at the expense of easing of a pressure wave.

In the reoriented material velocity of distribution of waves is maximal in a direction of axis Z, therefore the pressure wave loses energy only on fracture of a material and being reflected from a back surface by an intensive unloading wave breaking a material.

4.3. Numerical modeling of deformation and fracture of the composite spaced targets under impact

This paper presents a comparative analysis of fracture in monolithic and spaced barriers during high-velocity interaction with compact projectiles. The material of barriers is orthotropic organoplastic with a high degree of anisotropy of elastic and strength properties. We investigate the fracture efficiency of the protective properties of monolithic and spaced barriers depending on the orientation of anisotropic material properties in a range of velocities of impact from 750 to 3000 m/s.

At high-velocity impact on spaced designs the defining role in the fracture of the projectile and barriers is played by shock wave processes. As a result of the extension of these processes there is a fracture of the projectile and thin screens protecting the main design. Thin screens are effective for protection of space vehicles against particles of small space debris, moving at velocities more than 3 km/s at which intensive fracture of particles begins. At interaction with more massive particles with velocities not over 3 km/s thin screens are not so effective. In space similar situations are possible at the interaction of devices with space debris on catching up courses. At velocities of impact in a range not over 3 km/s an important role belongs to the strength characteristics of materials. Earlier conducted researches of the fracture of the spaced barriers made from isotropic metal materials have shown that their efficiency, in comparison with efficiency of the monolithic barrier, increases with increase in velocity of interaction. Now for manufacturing of aircrafts elements the various types of composite materials having a high degree of anisotropy of the elastic and strength properties are used, and without such property as anisotropy it is impossible to describe and predict the behavior of design elements and of the design as a whole (Radchenko & Radchenko, 2011).

On fig. 12, 13 computational configurations of the aluminum projectile, of the monolithic barrier and also of the two-layer and three-layer spaced barriers made from anisotropic organoplastic for velocities of impact from 750 to 2000 m/s accordingly are presented.

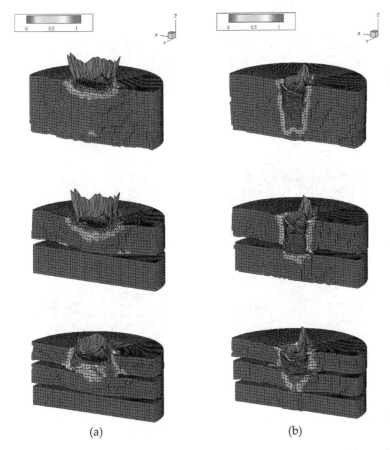

(a)   (b)

Figure 12. Calculated configurations of cooperating bodies and isolines of relative volume of fractures (V/V_0) in targets. a: υ_0 =750 m/s, β =0°, b: υ_0 =750 m/s, β =90°; t=40 μsec

On fig. 12a, fig. 13b barriers made from an initial material are given, on a fig. 12b, fig. 13b there are barriers made from reoriented material. It is necessary to explain that values of V

and V_0 are given in node of a mesh: V is a volume of incorporating elements of node in which the fracture condition was satisfied; V_0 is the total volume of the elements containing this node. Value of 1 for V/V_0 corresponds to the full fracture of the material in node of a mesh. As researches showed (Radchenko et al., 1999), a picture of the fracture observing in barriers depends on orientation of a material properties in relation to the direction of action of external loading, in this case to an impact direction. As a result the dynamic nature of fractures in a barrier defines an efficiency of its protective properties.

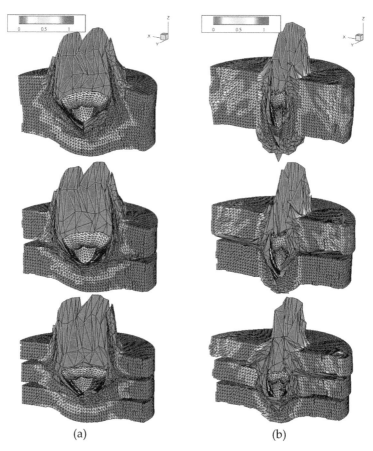

(a)   (b)

Figure 13. Calculated configurations of cooperating bodies and isolines of relative volume of fractures (V/V_0) in targets. a: υ_0 =2000 m/s, β =0°, b: υ_0 =2000 m/s, β =90°; t=40 μsec

The fracture of barriers material begins in a compression wave ($e_{kk} \leq 0$), initiated at the moment of impact. Fracture development in barriers depends not only on value of the strength material characteristics, but also on velocities of propagation of compression and unloading waves.

In barriers made from the reoriented material (fig. 12b, fig. 13b) at the expense of greater velocity of wave propagation in an impact direction (along the axis Z) and smaller value of strength on pressure in this direction, the fracture occurring in a compression wave gets deeply on a thickness of a barrier. The unloading waves moving from free surfaces, reaching areas in which the material was already weakened at pressure, completely destroy it. In this case the projectile has before itself an extended area of the fractured material, which doesn't resist to a projectile introduction ($\sigma_{ij} = 0$).

In barriers made from an initial material (fig. 12a, fig. 13a) we have other picture: the areas of fracture realized in a compression wave, have more extended sizes in directions which are perpendicular to an impact direction. Before the projectile the area of not fractured material providing greater resistance to the projectile introduction remains.

The dynamics of fracture in barriers is estimated fully on fig. 14, which shows changes in time of the total (on a barrier) relative volume of fractures at pressure (V_p/V_0) or tension (V_t/V_0) (V_0 is total volume of barrier) for various initial velocities of impact. Curves with V_p characterize change in time of relative volume of the material destroyed in conditions of pressure ($e_{kk} \leq 0$) and keeping resistance to loading only on pressure, for monolithic and spaced (two- or three-layer) barriers. Curves with V_t characterize change in time of relative volume completely fractured material ($e_{kk} > 0$), not resisting to loading ($\sigma_{ij} = 0$). As the material fracture begins in a compression wave, at the beginning of the process volume of the fractured material being in a condition of pressure, is more than volume of the fractured material at tension. Eventually volume of the material, fractured at tension, increases at simultaneous reduction of volume of the material fractured at pressure. It is caused by influence of unloading waves, extending from free surfaces of the projectile and barriers. If for projectile velocity of 750 m/s the volume of completely fractured material ($e_{kk} > 0$) in the monolithic barrier was the smallest, with increase in impact velocity the picture changes — the volume of the fractured material in a monolithic barrier begins to exceed the corresponding values for the spaced barriers.

The comparative analysis of protective properties of the monolithic and spaced barriers made from an initial material can be spent analogically on dependences of velocity of the center of the projectile weights from time, resulted on a fig. 15 for various orientations of a material properties. For an initial orientation of a material properties at velocities of impact to 2000 m/s the more intensive braking of the projectile on a monolithic barrier occurs (fig. 15a, fig. 15b, fig. 15c). And at velocity of 750 m/s the through penetration for the monolithic and for the spaced barriers is absent. In this case during all the process of the interaction the projectile velocity falls more at interaction with the monolithic barrier. Following on efficiency is the spaced barrier made from two plates, and the least effective is a barrier made from three plates.

With increase in initial velocity of impact the picture changes. For the impact velocity of 1500 m/s the monolithic barrier is still more effective – braking of the projectile on it occurs more intensively. Post-penetration velocity of the projectile after penetration of the monolithic barrier makes 50 m/s, after penetration of the spaced barriers – 150 m/s, but for the initial velocity of 1500 m/s in 36 μsec after the impact start the projectile velocities for cases of two-layer and three-layer spaced designs are leveled. The increase in velocity to 3000 m/s leads to that after 26 μsec after the process start the projectile begins to brake more intensively on the spaced targets, thus post-penetration velocity after penetration of the monolithic barrier is already on 10–15% above, than after penetration of the spaced barriers. For the reoriented material the tendency of increase of the spaced designs efficiency remains – with growth of velocity of interaction the difference in post-penetration velocities of the projectile after penetration of the monolithic and the spaced barriers decreases.

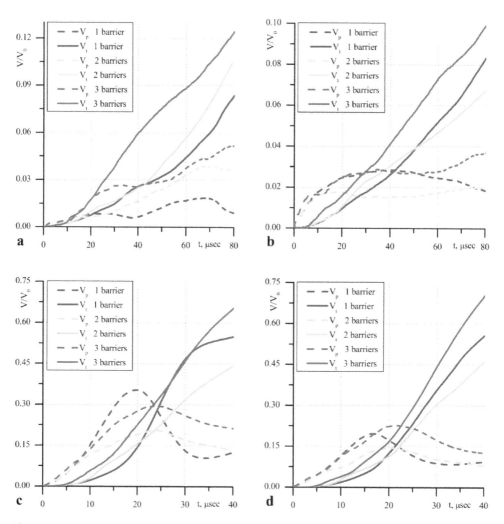

Figure 14. Relative volume of fractures at pressure and tension.

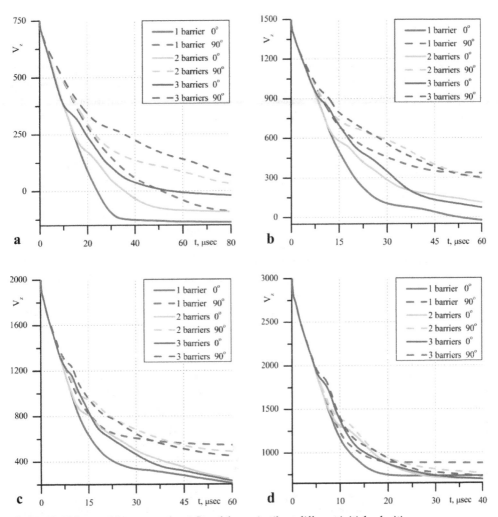

Figure 15. Velocity of the center of weights of the projectile at different initial velocities.

4.4. Numerical analysis of impact interaction of two anisotropic solids

Let's consider normal impact of orthotropic projectile with diameter 15 mm and length 75 mm and orthotropic barrier with thickness 60 mm with initial velocity $v_0 = 700$ м/c along axis Z (Radchenko & Radchenko, 2010). Orientation of properties of orthotropic material changes by turn of axes of symmetry of an initial material round an axis OY on a corner β, counted from a positive direction of an axis Z. Following cases of orientation of properties of a material of a projectile and a barrier are investigated: 0–0, 0–90, 90–0, and 90–90. Here numbers correspond to values β for a material of a projectile and a barrier in degrees, accordingly. Cases when the top and bottom half of a projectile and a barrier consist of materials reoriented on 90° are also considered (conditionally these cases are designated as 0–90 1/2 and 45–135 1/2).

On fig. 16 settlement configurations of the projectile and a barrier and as are presented an isolines of relative volume of fracture for various cases of orientation of properties of a material. The fig. 16a corresponds to a case (0–90 1/2) when the lower half of projectile and a barrier consists of an initial material in which the greatest strength characteristics of tension are oriented in an axis direction X, perpendicular to a impact direction, and in the upper half of projectile and a barrier of property of a material are reoriented on 90° by turn concerning an axis Y. The Fig. 16b corresponds to a case (45–135 1/2) when the lower half of projectile and a barrier consist of a material in which maximum strength characteristics of tension are oriented at an angle 45° to an axis X in a plane ZX, and in the upper halves properties of a material are reoriented by turn on 90° concerning an axis Y. In the first case (Fig. 16a), besides fracture of a head part, fracture of the projectile on boundary of materials with various orientation of properties at reaching by a wave of compression of the upper half of projectile (in which the material has the minimum strength on compression in an axis direction Z), that leads to its sharing on two parts is observed. In the second case (Fig. 16b) because axes of symmetry of materials of the projectile and a barrier do not coincide with coordinate axises and, accordingly, with a direction of propagation of shock waves, fracture process in the projectile and a barrier passes not symmetrically. In the upper half of barrier the fracture area is oriented in a direction of the minimum values of strength on compression (β=135°). In the projectile the fracture area as is formed on boundary of section of materials with various orientation of elastic and strength properties, but in this case fracture extends from a side surface of the projectile at an angle 135° and volume of fracture in a head part of the projectile much more.

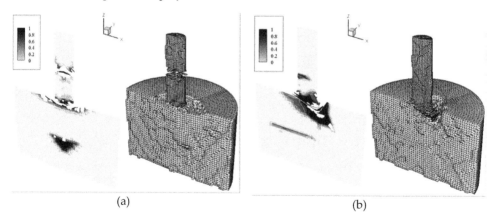

(a)   (b)

Figure 16. Configurations of interacting bodies and allocation of isolines of relative volume of fracture. υ_0 =700 m/s, t =36 μsec.

Dynamics of fracture in the projectile is illustrated by fig. 17 where changes in time of relative volume of the fracture derivated in the conditions of compression and tension are resulted. The analysis of curves on Fig. 17 allows to draw a conclusion that for all considered cases of interaction the least levels of irreversible fracture (Fig. 17b) are realized in the projectile from a material with orientation of axes of symmetry $\beta = 0°$. It is caused by that in this case the material of the projectile has the greatest strength properties on compression in an axis direction Z. Therefore the fracture at the initial stage of process of interaction in a wave of compression (fig. 17a), extending on length of the projectile, are minimal. For other cases of orientation of properties of a material of the projectile the volume of the fractured material in a wave of compression is more, in this connection the most part of a material of the projectile appears weakened, and further does not offer resistance at occurrence of tension powers. Curves on fig. 17a have a maximum in a range from 16 to 20 μsec, and then the volume of the material fracture at compression, decreases, after that instant the volume of irreversible fracture of a material increases at tension.

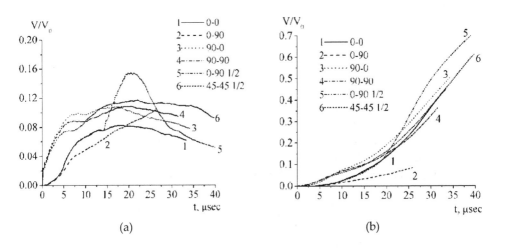

(a)   (b)

Figure 17. Change in time of relative volume of fracture in the projectile at compression (a) and tension (b) for various orientations of properties of a material.

Penetration power of the projectile depending on orientation of properties of a material at its interaction with various barriers can be estimated on the curves characterizing change in time of velocity of centre of mass the projectile (fig. 18). The most intensive braking of the projectile is observed for a case of orientation of properties 0–0 (a curve 1) and 0–90 1/2 (a curve 5). In the first of these two cases of property of a material of the projectile and a barrier are oriented equally and correspond to a case of maximum strength on compression in a direction of impact (axis Z), that stipulates high firmness of a barrier to impact. In the second case the projectile has extensive fracture and is divided on two parts (fig. 16a) that leads to loss of its penetration power.

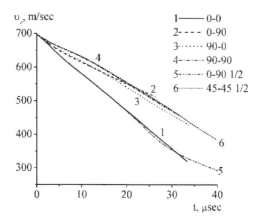

Figure 18. Change in time of velocity of centre of mass of the projectile.

5. Conclusion

1. The offered model allows to describe adequately main laws of the fracture processes of anisotropic materials under dynamic loads. The carried out researches have shown, that anisotropy of properties is the essential factor which is necessary for taking into account for the adequate description and the prediction of development of shock-wave processes and fracture in the materials under dynamic loadings. The influence of anisotropic properties orientation increases with decrease in the velocity interaction.

2. The qualitative and quantitative discrepancies in the fracture of isotropic and anisotropic materials under the dynamic loads are defined not only by strength parameters but either by the interaction of the compression and tension waves. Different velocities of waves propagation along the directions in anisotropic barriers provide the discharge of the impact wave and the narrowing in the fracture region.

3. The influence of hydrostatic pressure on the fracture of anisotropic materials under dynamic loads is shown. Full compression leads to a fragmentation of an orthotropic ball without causing the fracture of an isotropic ball. That is why adequately description of the dynamic behavior of anisotropic solids required accounting of the hydrostatic pressure.

4. It is established that at low-velocity impact formation and a direction of development of fracture zone in a target is defined by orientation of elastic and strength properties of an anisotropic material in relation to an impact direction. Depending on orientation of properties development of the conic cracks caused by combined action of tensile stresses in waves of unloading and at the expense of penetration of the projectile, or fracture of material in a pressure and unloading wave is probable.

5. A comparative analysis of the effectiveness of the protective properties of monolithic and spaced barriers from anisotropic materials for various cases of orientation of the

material properties is carried out. It is established that the effectiveness of spaced designs increases with the velocity of interaction and determined by the development of fracture in barriers, depending on the velocities of wave propagation and the orientation of the elastic and strength properties of anisotropic material with relationship to the direction of impact.

Orientation of properties of an anisotropic material of the projectile and a barrier essentially influences dynamics of fracture and on process of propagation of shock waves in the projectile and a barrier. The offered model of behavior of anisotropic materials at dynamic loads allow to carry out researches by definition of optimization of properties of a material of the projectile and a barrier at any orientation of axes of symmetry.

Author details

Andrey Radchenko
Tomsk State University of Architecture and Building, Russia

Pavel Radchenko
Institute of Strength Physics and Materials Science of SB RAS, Russia

6. References

Ashkenazi, E. K., & Ganov, A.V. (1980) *Anisotropy of the construction materials*, Leningrad

Johnson, G. R. Three-dimensional analysis of sliding surface during high velocity impact. (1977) *Journal of Applied Mechanics*, No. 6, pp. 771-773

Johnson, G.R. High Velocity Impact Calculations in Three Dimension. (1977) *Journal Applied Mechanics*, Vol. 3, p. 95–100

Kanel, G.I., Razorenov, S.V., Utkin, A.V., & Fortov, V.E. (1996) *Shock wave phenomena in condensed media*, Yanus-K, Moscow

Radchenko, A. V., Kobenko, S.V., Marcenuk, I.N., Khorev, I.E., Kanel, G.I., & Fortov, V.E. (1999) Research on features of behaviour of isotropic and anisotropic materials under impact. *International Journal of Impact Engineering*, Vol. 23, No. 1, pp. 745–756

Radchenko, A., Radchenko, P., Tuch, E., Krivosheina, M., & Kobenko, S. (2012) Comparison of Application of Various Strength Criteria on Modeling of Behavior of Composite Materials at Impact. *Journal of Material Science and Engineering A*, Vol. 2, No. 1, pp. 112–120

Radchenko, A.V., & Radchenko, P.A. (2010) Numerical analysis of impact of two anisotropic solids. *Journal of Materials Science and Engineering*, Vol. 4, No. 3, pp. 74–79

Radchenko, A.V., & Radchenko, P.A. (2011) Numerical modeling of development of fracture in anisotropic composite materials at low-velocity loading. *Journal of Materials Science*, Vol. 46, No. 8, pp. 2720–2725

Tsai, S.W., & Wu, E.M. (1971) A General Theory of Strength for Anisotropic Materials. *Journal of Composite Materials,* Vol. 5, No. 1, pp. 58–80

Wilkins, M.L., & Guinan, M.W. (1973) Impact of cylinders on a rigid boundary. *Journal of Applied Physics,* Vol. 44, No. 3, pp. 1200–1206

Numerical Modelling of Damage Evolution and Failure Behavior of Continuous Fiber Reinforced Composites

F. Wang and J. Q. Zhang

Additional information is available at the end of the chapter

1. Introduction

Demands on high performance materials for use in a spectrum of structural and non-structural applications are increasing to the point that monolithic materials cannot fully satisfy multifunctional requirements. One approach that has emerged is to develop advanced fiber composites whose properties are tailored for wide applications in engineering, such as turbine generators and ultrasonic aircrafts (Miracle, 2005). Composite materials are composed of reinforcements and matrices, in which mechanical behaviors can be seriously affected by their micro-structures. Therefore, it becomes necessary to deal with these materials from a micromechanical view point. Most importantly, an urgent task in the field of fiber-reinforced composites is to develop a relationship between the material structure and its performance under more severe loads and environments.

Damage and failure in fiber-reinforced composites evolves at several different length scales. At the smallest scale, pre-existing defects in fibers grow. Due to statistical distribution of such defect, the fiber strength exhibits large variability. Therefore, as increasing load is applied to the composite, the weakest fiber will first break. The loads carried by the broken fiber are redistributed among the remaining unbroken fibers and matrix as determined by the constitutive response of the fibers, matrix and interface. And this then causes other fibers near the failure site to fail and thus shed further load to intact fibers. Consequently, the failure of fibrous composites goes through a very complicated damage evolution, which is a combination of fiber fracture, matrix deformation and interfacial debonding and slipping around the fiber breaks, before it reaches ultimate failure. It is obvious that the connection between the microstructural scale and the macroscopic scale is nontrivial and involves mechanics, stochastics, and volume scaling (Curtin, 1999).

Local damage evolution at the scale of fiber diameter in composites under loading determines its fracture toughness, strength, and eventually lifetime. The mechanical behavior of such a composite depends on the evolution of multi-damages throughout the application of loading, and then modelling composites undergoing progressive damage becomes a complex procedure due to many mechanisms involved (Mishnaevsky & Brøndsted, 2008). The development of a micromechanical damage model has to take different aspects into account such as (i) the proposition of a micromechanics-based approach that describes the influence of the damage variable on material properties (Blassiau et al., 2008), (ii) the definition of a pertinent damage variable and its law of evolution (Kruch et al., 2006), and (iii) the use of an appropriate and efficient experimental technique for the evaluation of damage, such as neutron diffraction (Hanan et al., 2005) and acoustic emission (Bussiba et al., 2008), etc.

Reliability concerns in utilizing fibrous composites in structural application have motivated the development of many numerical and analytical failure models in the presence of multidamages. Different from the phenomenological approaches based upon the macroscopic level, the progressive model is needed to consider local damage mechanisms, such as fiber breakage, matrix deformation, interfacial debonding, etc (Kabir et al., 2006) and predict the dominant failure modes. This method seems to be more accurate but computationally complicated because it accounts for many failure mechanisms and is also related to damage accumulation correlated with material properties degradation. Recently computational micromechanics is also emerging as an accurate tool to study the mechanical behaviors of composites because of the sophistication of the modeling tools and the ever-increasing power of digital computers (González, 2004). Within the framework, the macroscopic properties of a composite can be obtained by means of numerical simulation of the deformation and failure of the microstructure (Xia et al., 2001; F. Zhang et al., 2009).

In our previous work, the local cyclic shear plasticity of the interface around a broken fiber in ductile matrix composites under the in-phase and out-of-phase thermo-mechanical fatigue loads was analyzed by using the single-fiber shear-lag model (Zhang et al., 2002). A multifiber shear-lag model including matrix tensile modulus based upon an influence function superimposition technique was developed to simulate the nonlinear stress-strain response and the progressive failure of continuous fiber reinforced metal matrix composites under static tensile loading (Zhang & Wang, 2009) as well as thermomechanical fatigue loading (Zhang & Wang, 2010). This chapter will summarize our work and be organized as follows: an analytical model of the fibrous composite will be presented in Section II, in which fiber strength statistics, matrix behavior, and interfacial mechanics are explained in details. In Section III, we will introduce an influence function superimposition technique to derive stress profiles for any configuration of breaks, by considering local matrix plasticity and interface yield. In Section IV, numerical models combined with Monte-Carlo method will be developed to simulate progressive damage. In Section V, we will investigate failure behaviors of continuous fiber reinforced composites under cyclically thermomechanical loading. Finally, we will discuss limitations of the existing models, and aspects of the existing theories that require improvement.

2. Analytical model for multiple damage events

In the real multifiber composites of practical interest, the evolution of the fiber fragmentation during loading is, in principle, different because each individual fiber experiences a nonuniform stress due to the uniform applied stress plus stresses transferred from other broken fibers in the composite. The evolution of fiber damage thus depends crucially on the nature of the load transfer from broken or slipping fibers to unbroken fibers.

Consider an infinite, two-dimensional (2D) unidirectional composites reinforced with parallel, evenly-spaced fibers embedded in matrix material, in which the fibers and matrix regions are numbered in the serial number, shown in Fig. 1. The lamina is loaded in simple tension along the fiber direction. The width of fibers is D, and the width of each matrix region W can be related to the fiber volume fraction, V_f (Zhang et al., 2002).

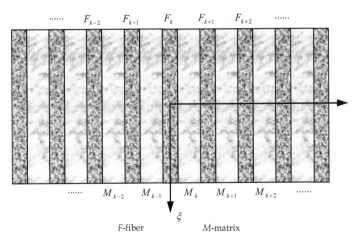

Figure 1. Schematic of the 2D fiber-matrix arrangement.

In many typical applications, fibrous composites are subjected to a cyclic mechanical loading with a superimposed variation in temperature. This type loading condition is referred to as thermomechanical fatigue (TMF) and can be regarded as one of the most severe types because the significant difference in the coefficient of the thermal expansion between the matrix and the fiber causes high thermal stresses and stress amplitudes raising irreversible deformation in the composites (Mall & Schubbe, 1994). In-phase and out-of-phase thermomechanical fatigue (TMF) loads are illustrated in Fig. 2. The tensile stress applied to composites varies between (σ_∞) min and (σ_∞) max while the temperature changes between T_{max} and T_{min}. For convenience, the state 'A' is used to represent the state of TMF loads with the maximum applied stress, and the state 'B' denotes the state of TMF load with the minimum mechanical load. It is clearly illustrated that in-phase conditions subject the composite to high stresses at hot temperatures and low stresses at low temperatures. Conversely, out-of-phase TMF conditions subject the composite to high stresses at low temperatures and vice versa since the peaks in the waveforms are 180° apart (Williams & Pindera, 1995).

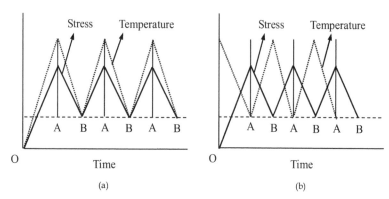

(a)   (b)

Figure 2. Stress and temperature for, (a) in-phase, (b) out-of-phase thermomechanical cycling.

Since the fiber strength exhibits large variability due to statistic distribution of defect, the weakest fiber break usually takes place at very early loading stage. After some loading level there are N fiber breaks occurred in the composite. The coordinates of the n-th fiber break, which locates in the k_n-th fiber with vertical position ξ_n, are given by (k_n, ξ_n) $(n = 1, 2,..., N)$.

If the interfacial strength is higher than the matrix yield stress in shear, the plastic deformation of the matrix will occur before fiber/matrix debonding (Beyerlein & Phoenix, 1996). This was observed in ductile resin and some metal matrix composites with a strong interface. Some experiments show that the microdamage and deformation modes include the fiber breakage, fiber pullout, debonding, and local plasticity around fiber breaks (Liu & He, 2001).

In order to analyze the complex combination of microdamage and deformation, we may propose a micromechanical model, shown in Fig. 3., in which a broken fiber, accompanying with its (reserve) tensile yielding matrix and its (reverse) shear yielding interface, and its debonding interface is called as a damage-plasticity event. Essentially the nonlinear cyclic behavior and thermomechanical fatigue failure of the composite is the result of the interactions among these damage-plasticity events under TMF loads.

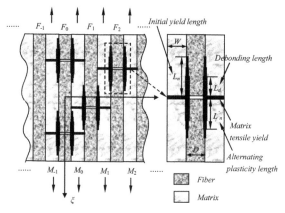

Figure 3. Multi-damage modes in composites.

In the loading phase the local interface shear yield and the matrix tensile yield around the fiber break could take place due to stress concentration as the load increases. The initial interface yield length caused during the loading phase is represented by, $2L_n$, associated with the n-th fiber break. For fatigue case, when the load is released from state 'A' to state 'B' in the unloading phase, the yielded interface and the yielded matrix around the fiber breaks start the elastic unloading. The interface shear stress at broken sites relaxes faster than that at any other positions. Thus, the shear stress of broken sites will change sign. Subsequently, there are two possible cases: the complete elastic unloading and the reverse shear plastic yielding within a certain length, depending on the load range. If the load range is not such large that the shear stress of the broken site does not reach the reverse yield stress before state 'B', the unloading is completely elastic. In this case the unloading-reloading will be totally elastic, that is, shakedown occurs. If the load range is sufficiently large, the interface shear stress around the broken sites will change sign and will reach the reverse yield stress at some moment before the load reached state 'B'. The interface plasticity is modeled by the elastic, perfectly plastic shear stress-strain relation, ignoring the effect of the interface tensile stress, that is,

$$\tau = \begin{cases} G_c \gamma & |\tau| < \tau_s \\ \tau_s(T) & \text{yielded} \end{cases}, \tag{1}$$

where G_c and τ_s are the shear modulus of the interface and the yield shear stress, respectively; τ and γ denote the interfacial shear stress and shear strain, respectively. This interface constitutive law allows the interface to undergo the cyclic plasticity under TMF loading. The yield shear stress is related to the temperature by

$$\tau_s = c_1 + c_2 \left[1 - \exp(c_3 T) \right], \tag{2}$$

where c_1, c_2 and c_3 are material constants.

With continuous mechanical unloading of composite, the reverse plastic yield of the interface will extend forward from the fiber break tip. Up to the loading moment 'B', the reverse plastic yield has occurred within a certain length, $2L'_n$. With the repeated loading and unloading processes, the cyclic interface shear plasticity takes place within the length, $2L'_n$, leading to debonding. In other words, the interface is debonded in the length of $2L_d$ because of cyclic plastic shear strain accumulation. It must be pointed out that material parameters in cyclic interface shear plasticity and debonding growth are difficult to measure directly. For the strong interface considered here, it is assumed that the interface shear yield is governed by the shear yield of matrix in a very thin layer. In other words, the material parameters of the interface are assumed to the same as those of the matrix.

Matrix tensile yield in the matrix regions neighboring with the fiber break sites could take place due to matrix tensile stress transferred from the tensile load released from the broken fiber. It is assumed that the M breaks of the N fiber breaks ($M \leq N$) cause the associated matrix tensile yielding. Each matrix tensile yield zone takes the shape of a narrow strip spanning the matrix width, and is colinear with the associated fiber break. The constitutive behavior of matrix in the yield zones obeys the uniaxial cyclic elastic, perfectly plastic stress-strain relation. The yield tensile stress σ_s is related to the yield shear stress τ_s by von Mises yield criterion, that is, $\sigma_s = \sqrt{3}\tau_s$.

3. Stress profiles for multiple damage events

Accurate prediction of the mechanical properties of the composite materials requires detailed knowledge of the micromechanical stress state in and around broken fibers as a function of the constituent material properties (Xia et al., 2002). Stress transfer between fiber and matrix is one of fundamental issues for a composite system (Hanan et al., 2003; Beyerlein & Landis, 1999). In order to obtain solutions for deformation in the composites containing multiple damage events, a theoretical model under the framework of 2D shear-lag arguments will be used to derive stress profiles for any configuration of breaks in the presence of matrix tensile yield and interface shear yield. It is assumed that the tensile load in a fiber-reinforced composite loaded in the fiber direction is mainly carried by the fibers and matrix regions, where the load between fiber and matrix is transferred through the fiber/matrix interface shear stress. The transverse deformation in the fibers and matrix regions is neglected. The interface is assumed to be in as state of shear behavior as controlled by the axial displacements of the fiber and matrix materials. Shear-lag analysis of stresses and deformations, involving the cyclic thermoplasticity, must be carefully carried out by applying TMF loads to the composites incrementally.

3.1. Stress solution in loading phase

Let us consider the TMF loading phase. As the thermomechanical loads increase, N fiber breaks have appeared where each fiber break is accompanied by the interface yielding of length, $2L_n$, and where M breaks ($M \leq N$) cause the associated matrix tensile yielding simultaneously. The interfacial shear stress within the yield zone $0 \leq |\xi - \xi_n| \leq L_n/l_c$ associated with the n-th fiber-break has to be equal to the yield shear stress, it follows

$$\tau\left(k_n, \xi\right) = \tau_s, \text{ for } x = x_n + L_n / l_c \text{ and } n = 1,2,3,\ldots,N, \tag{3}$$

where τ (k_n, ξ) is the interface shear stress at the location, (k_n, ξ). Since the shear stress is constant (τ_s) within the interface yield region, the fiber stress can be obtained simply from the equilibrium condition of fiber, that is,

$$\sigma^f\left(k_n, \xi\right) = \frac{2\tau_s}{D}\left|l_c\left(\xi - \xi_n\right)\right|, \text{ for } 0 \leq |x - x_n| \leq L_n / l_c \text{ and } n = 1,2,3,\ldots,N, \tag{4}$$

where σ^f (k_n, ξ) is fiber stress at the location, (k_n, ξ). Matrix tensile stress in the narrow matrix yield zone is

$$\sigma^m\left(k_t, \xi_t\right) = \sigma_s, \ t = 1,2,3,\ldots,M, \tag{5}$$

In order to obtain solutions for fiber stress and stress in matrix outside the yield zones the superposition method (Beyerlein & Phoenix, 1996; Landis et al., 2000) has been further extended in our previous work (Zhang & Wang, 2009, 2010). Original superposition method of Beyerlein excludes the axial load-bearing capacity of matrix, which is suitable for polymer matrix composites. Our extended superposition method includes the axial load-bearing capacity of matrix, which is applicable to fiber reinforced metal matrix composites. The fundamental concepts behind *influence superimposition techniques* are to obtain analytical solutions for the influence of a 'unit' failure on the stress and displacement fields and then use a weighted superposition technique to obtain the solution for the full problem. Indeed, the units solutions for a single fiber-break and a single matrix-break, called the influence functions, are used as building blocks for the solutions for fiber stresses and matrix stresses in the composites.

In the method, the narrow matrix yield zones is modeled as fictitious matrix breaks with surface tensile traction σ_s, and the interface yield length is considered as a fictitious debonding crack with surface shear traction τ_s. The M fictitious matrix breaks (matrix yield zones) can be decomposed into the M problems of single matrix-break with opening displacements, w_t^m ($t = 1,2,3,\ldots,M$), that are yet to be determined by stress condition, eq.(5). For an interface yield length $0 \leq |\xi - \xi_n| \leq L_n/l_c$ associated with the n-th fiber break, the fiber segment with the known fiber stress given by eq. (4) can be modeled as a continuous distribution of fictitious fiber break with non-zero opening displacement $w_n^f(\xi)$ which is to be determined through the stress conditions, eqs.(3-4).

Therefore, the axial stresses in fibers and matrix regions and the interfacial shear stress, σ^f (k, ξ), σ^m (k, ξ) and τ (k, ξ) at any position (k, ξ) can be obtained by using the superposition method as follows,

$$\sigma^f\left(k,\xi\right)=\overline{\sigma}_f+\sum_{j=1}^{N}\int_{\xi_j-L_j/l_c}^{\xi_j+L_j/l_c}w_j^f\left(\xi'\right)p_{k-k_j}^f\left(\xi-\xi'\right)d\xi'+\sum_{i=1}^{M}w_i^m q_{k-k_i}^f\left(\xi-\xi_i\right), \tag{6}$$

$$\sigma^m\left(k,\xi\right)=\overline{\sigma}_m+\sum_{j=1}^{N}\int_{\xi_j-L_j/l_c}^{\xi_j+L_j/l_c}w_j^f\left(\xi'\right)p_{k-k_j}^m\left(\xi-\xi'\right)d\xi'+\sum_{i=1}^{M}w_i^m q_{k-k_i}^m\left(\xi-\xi_i\right), \tag{7}$$

$$\tau\left(k,\xi\right)=\sum_{j=1}^{N}\int_{\xi_j-L_j/l_c}^{\xi_j+L_j/l_c}w_j^f\left(\xi'\right)s_{k-k_j}\left(\xi-\xi'\right)d\xi'+\sum_{i=1}^{M}w_i^m t_{k-k_i}\left(\xi-\xi_i\right), \tag{8}$$

where $\overline{\sigma}_f$ and $\overline{\sigma}_m$ denote the remote stress in fiber and matrix, respectively. The influence functions $p_{k-k_j}^f\left(\xi-\xi'\right)$, $p_{k-k_j}^m\left(\xi-\xi'\right)$ and $s_{k-k_j}\left(\xi-\xi'\right)$ represents fiber stress, matrix stress and interface shear stress at position (k, ξ) due to a unit opening displacement applied at the fiber-break with position (k_j, ξ'), respectively. The influence functions $q_{k-k_i}^f\left(\xi-\xi_i\right)$, $q_{k-k_i}^m\left(\xi-\xi_i\right)$ and $t_{k-k_i}\left(\xi-\xi_i\right)$ stand for fiber stress, matrix stress and interface shear stress at position (k, ξ) due to a unit opening displacement applied at the fictitious matrix break (k_i,

ξ), respectively. The expressions for the influence functions are given in the appendix. The undetermined coefficient functions w_j^f ($j = 1, 2,..., N$), the unknown coefficient constant w_i^m ($i = 1, 2,..., M$) and the interface yield lengths L_j ($j = 1, 2,..., N$) can be obtained by the ($2N+M$) stress conditions given by eqs. (3-5). Substitution of eqs. (6-8) into eqs. (3-5) gives

$$\sum_{j=1}^{N} \int_{\xi_j - L_j/l_c}^{\xi_j + L_j/l_c} w_j^f\left(\xi'\right) s_{k_n - k_j}\left(\xi_n + L_n/l_c - \xi'\right) d\xi' + \sum_{i=1}^{M} w_i^m t_{k_n - k_i}\left(\xi_n + L_n/l_c - \xi_i\right) = \tau_s, \text{ for } n = 1, 2, 3,..., N, \quad (9)$$

$$\bar{\sigma}_f + \sum_{j=1}^{N} \int_{\xi_j - L_j/l_c}^{\xi_j + L_j/l_c} w_j^f\left(\xi'\right) p_{k_n - k_j}^f\left(\xi - \xi'\right) d\xi' + \sum_{i=1}^{M} w_i^m q_{k_n - k_i}^f\left(\xi - \xi_i\right) = \frac{2\tau_s}{D}\left|l_c\left(\xi - \xi_n\right)\right|,$$
$$\text{for } 0 \leq |x - x_n| \leq L_n/l_c \text{ and } n = 1, 2, 3,..., N, \quad (10)$$

$$\bar{\sigma}_m + \sum_{j=1}^{N} \int_{\xi_j - L_j/l_c}^{\xi_j + L_j/l_c} w_j^f\left(\xi'\right) p_{k_t - k_j}^m\left(\xi_t - \xi'\right) d\xi' + \sum_{i=1}^{M} w_i^m q_{k_t - k_i}^m\left(\xi_t - \xi_i\right) = \sigma_s,$$
$$\text{for } t = 1, 2, 3,..., M, \quad (11)$$

where $l_c = \sqrt{\dfrac{WE_m}{K}}$; E_m are the Young's modulus of the matrix and K is the shear-lag parameter associated with fiber/matrix interfaces. By solving the integral eqs. (9-11) one can obtain the shear yield length, L_n, the coefficient functions w_j^f ($j = 1, 2,..., N$) and coefficients w_i^m ($i = 1, 2,..., M$). Physically w_j^f and w_i^m represent the fiber displacement and the elongation of the narrow matrix yield zone.

It is worthy to emphasize that the stress analysis method mentioned above is exact under the framework of shear-lag argument. In that sense, moreover, the mechanical interactions between the multiple damage-plasticity events have been fully and exactly taken into account since the governing equations of shear lag model, the boundary conditions and the plastic yield conditions have been fully satisfied.

3.2. Stress solution in unloading phase

There are two possible cases. If the unloading is completely elastic, the solutions for the stresses and the displacements at the state 'B' can be attained by superposing the elastic solutions between A and B onto the solutions of the state 'A'. Otherwise, if the load range is sufficiently large, the local reverse yield of the interface and matrix will takes place before the thermomechanical load reaches state 'B'. We consider an unloading increment, during which the interface reverse yielding occurs in N' ($N' \leq N$) lengths, $2L'_n$ ($n = 1, 2,..., N'$), and matrix reverse yielding takes place in M' ($M' \leq M$) narrow matrix zones. The interfacial shear stress within the reverse yield zone $0 \leq |\xi - \xi_n| \leq L'_n/l_c$ has to be equal to the reverse yield shear stress, i.e.

$$\tau\left(k_n, \xi\right) = -\tau_s, \text{ for } 0 \leq |x - x_n| \leq L'_n/l_c \text{ and } n = 1, 2, 3,..., N', \quad (12)$$

Within the interface reverse yield region the fiber stress can be easily obtained from the equilibrium condition of fiber as follows

$$\sigma^f\left(k_n,\xi\right)=-\frac{2\tau_s}{D}\left|l_c\left(\xi-\xi_n\right)\right|, \text{ for } 0\leq\left|x-x_n\right|\leq L'_n/l_c \text{ and } n=1,2,3,\ldots,N', \tag{13}$$

Matrix stress in the narrow matrix reverse yield zone is

$$\sigma^m\left(k_t,\xi_t\right)=-\sigma_s \text{ , } t=1,2,3,\ldots,M', \tag{14}$$

The incremental stress solutions for fibers and matrix regions in composite for the thermomechanical unloading increment can be obtained by using the superposition method mentioned in the previous section. The solutions for the total stresses after the thermomechanical unloading increment, then, can be obtained by superposing the incremental solutions onto the solutions before the unloading increment, that is,

$$\sigma^f\left(k,\xi\right)=\sigma_0^f\left(k,\xi\right)+\sum_{j=1}^{N'}\int_{\xi_j-L'_j/l_c}^{\xi_j+L'_j/l_c}\overline{w}_j^f\left(\xi'\right)p_{k-k_j}^f\left(\xi-\xi'\right)d\xi'+\sum_{i=1}^{M'}\overline{w}_i^m q_{k-k_i}^f\left(\xi-\xi_i\right), \tag{15}$$

$$\sigma^m\left(k,\xi\right)=\sigma_0^m\left(k,\xi\right)+\sum_{j=1}^{N'}\int_{\xi_j-L'_j/l_c}^{\xi_j+L'_j/l_c}\overline{w}_j^f\left(\xi'\right)p_{k-k_j}^m\left(\xi-\xi'\right)d\xi'+\sum_{i=1}^{M'}\overline{w}_i^m q_{k-k_i}^m\left(\xi-\xi_i\right), \tag{16}$$

$$\tau\left(k,\xi\right)=\tau_0\left(k,\xi\right)+\sum_{j=1}^{N'}\int_{\xi_j-L'_j/l_c}^{\xi_j+L'_j/l_c}\overline{w}_j^f\left(\xi'\right)s_{k-k_j}\left(\xi-\xi'\right)d\xi'+\sum_{i=1}^{M'}\overline{w}_i^m t_{k-k_i}\left(\xi-\xi_i\right), \tag{17}$$

where the coefficients $\overline{w}_j^f$ ($j=1, 2,\ldots, N'$) and $\overline{w}_i^m$ ($i=1, 2,\ldots, M'$) are to be determined which belong to the incremental unloading. The subscript '0' represents the state before an incremental unloading. The undetermined interface reverse yield lengths $2L'_n$ are governed by the interfacial shear stress at the end of reverse yield zone, $\overline{\xi}_n=\xi_n+L'_n/l_c$, and it follows from eqs. (12) and (17) that

$$\tau_0\left(k_n,\overline{\xi}_n\right)+\sum_{j=1}^{N'}\int_{\xi_j-L'_j/l_c}^{\xi_j+L'_j/l_c}\overline{w}_j^f\left(\xi'\right)s_{k_n-k_j}\left(\xi_n+L'_n/l_c-\xi'\right)d\xi'+\sum_{i=1}^{M'}\overline{w}_i^m t_{k_n-k_i}\left(\xi_n+L'_n/l_c-\xi_i\right)=-\tau_s, \tag{18}$$

for $n=1,2,3,\ldots,N'$

Substitution of eqs. (15-16) into eqs. (13-14) leads to

$$\sigma_0^f\left(k_n,\xi\right)+\sum_{j=1}^{N'}\int_{\xi_j-L'_j/l_c}^{\xi_j+L'_j/l_c}\overline{w}_j^f\left(\xi'\right)p_{k_n-k_j}^f\left(\xi-\xi'\right)d\xi'+\sum_{i=1}^{M'}\overline{w}_i^m q_{k_n-k_i}^f\left(\xi-\xi_i\right)=-\frac{2\tau_s}{D}\left|l_c\left(\xi-\xi_n\right)\right|, \tag{19}$$

for $0\leq\left|x-x_n\right|\leq L'_n/l_c$ and $n=1,2,3,\ldots,N'$

$$\sigma_0^m\left(k_t,\xi_t\right)+\sum_{j=1}^{N'}\int_{\xi_j-L'_j/l_c}^{\xi_j+L'_j/l_c}\overline{w}_j^f\left(\xi'\right)p_{k_t-k_j}^m\left(\xi_t-\xi'\right)d\xi'+\sum_{i=1}^{M'}\overline{w}_i^m q_{k_t-k_i}^m\left(\xi_t-\xi_i\right)=-\sigma_s, \text{ for } t=1,2,3,\ldots,M' \tag{20}$$

By solving the integral eqs. (18-20) one can obtain the coefficient functions $\overline{w}_j^f$ ($j = 1, 2,...,$ N'), coefficients $\overline{w}_i^m$ ($i = 1, 2,..., M'$) and the reverse shear yield lengths L'_n ($n = 1, 2,..., N'$).

3.3. Local cyclic plasticity and debonding

The cyclic shear plasticity of the interface will take place only if $L'_n > 0$. Otherwise there is no alternating plastic shear and interface undergoes cyclic elastic deformation over the whole length after initial yielding of length L_n, that is, shakedown. Since the elastic, perfect-plastic constitutive relation has been assumed for the interface shear and matrix, the solutions for all cycles are the same before the new debonding takes place. The alternating plastic shear strain range $\Delta\gamma_p = \gamma_p^A - \gamma_p^B$ can be derived from displacement solutions for loading and unloading condition. The total interface shear strain is evaluated by eq. (21a)

$$\gamma = 2\frac{u^m - u^f}{W + D}, \quad (a) ,$$

$$\gamma_p = \gamma - \frac{\tau}{G_c} \quad (b)$$

(21)

where u^f is the axial displacement of broken fiber and u^m is the axial displacement of matrix adjacent to the broken fiber. The plastic shear strain can be easily computed by eq. (21b).

Debonding is an important mechanism that stimulates stress redistribution in composites. Since plastic strain is dominant in the low cyclic fatigue regime (Llorca, 2002), the cyclic plastic shear strain range $\Delta\gamma_p$ may be used to predict the debonding (Zhang et al., 2002) along with Coffin-Manson fatigue equations $\Delta\gamma_p/2 = \gamma_f \cdot (2N_f)^c$, where N_f is the cycles to interface debonding under the constant shear plastic strain range $\Delta\gamma_p$, material constants γ_f and c stand for the fatigue ductility coefficient and the fatigue ductility exponent, respectively. The debonding growth law based on Coffin-Mason equations represented by plastic shear strain range provides a means to account for the stable growth of debonding as the number of cycles increases, which may be relevant for the strong interface and/or the low applied load where the static debonding criterion fails to predict a further growth of debonding. The parameters in the debonding growth law are taken to the same as matrix since we assumed that the interface shear yield is governed by the shear yield of matrix in a very thin layer. This criterion for debonding based upon the shear-lag model ignores the effect of the interface tensile stress. This simplification is reasonable for the growth-dominated fatigue failure since the shear stress contributes to its growth and propagation along the fiber length. However, the interface tensile stress component could be important for the cause of the initiation of fiber/matrix debonding. Therefore, this simplification would predict a delayed debonding, leading to an overestimated fatigue lifetime for the initiation-dominated fatigue failure (high load). Although the composite stress range and the temperature range for both the in-phase and out-of-phase TMF loading conditions do not change with cycles, the local plastic shear strain ranges will change, due to new debonding, after some number of cycles. Therefore, one needs a damage accumulation fatigue rule to describe the debonding due to local varying amplitude fatigue. For simplicity, we assume that the fatigue life for interface debonding is governed by the linear damage accumulation rule $\sum N_i/(N_f)_i = 1$, where N_i is the number of the applied loading cycle leading to the constant plastic strain range $(\Delta\gamma_p)_i$ and $(N_f)_i$ is the cycles to interface debonding for the constant shear plastic strain range $(\Delta\gamma_p)_i$. When the sum of the fractions from each step equals one, debonding is predicted (Liu, 2001)

3.4. Stress distribution due to debonding

After interfacial debonding, there can remain a residual shear sliding resistance across the fiber/matrix interface due to friction. For tractability, an assumption is that a constant frictional shear force, τ_f, governed by Coulomb's law, exists within the debonded interface of the length, L_d. In new loading phase after debonding, the interfacial shear stress within the interface debonding length and the shear yield zone has to be equal to

$$\tau\left(k_n,\xi\right)=\begin{cases}\tau_f & 0\leq\left|\xi-\xi_n\right|\leq L_d/l_c \\ \tau_s & L_d/l_c\leq\left|\xi-\xi_n\right|\leq\left(L_n-L_d\right)/l_c\end{cases} \quad,\text{ for } n=1,2,3,...,N' \quad (22)$$

where N denotes the number of broken fibers which are accompanied by interface debonding and yielding. According to the continuity conditions between debonding and yield segments $\sigma^f\big|_{z=L_d^-}=\sigma^f\big|_{z=L_d^+}$, and the equilibrium condition for the fiber, the fiber stress within these two length has the form

$$\sigma^f\left(k_n,\xi\right)=\begin{cases}2\tau_f\cdot\left|l_c\left(\xi-\xi_n\right)\right|/D & 0\leq\left|\xi-\xi_n\right|\leq L_d/l_c \\ 2\left(\tau_s\cdot\left|l_c\left(\xi-\xi_n\right)\right|+L_d\left(\tau_f-\tau_s\right)\right)/D & L_d/l_c\leq\left|\xi-\xi_n\right|\leq\left(L_n-L_d\right)/l_c\end{cases} \quad,\text{ for } n=1,2,3,...,N' \quad (23)$$

In subsequent TMF unloading, the reverse shear plasticity may occur in N' lengths $2L'_n$ ($n = 1,2,3,...,N'$). In this case the interface shear stress is given by

$$\tau\left(k_j,\xi\right)=\begin{cases}-\tau_f & 0\leq\left|\xi-\xi_j\right|\leq L_d/l_c \\ -\tau_s & L_d/l_c\leq\left|\xi-\xi_j\right|\leq\left(L'_j-L_d\right)/l_c\end{cases} \quad,\text{ for } j=1,2,3,...,N' \quad (24)$$

It follows for the fiber stress,

$$\sigma^f\left(k_j,\xi\right)=\begin{cases}-2\tau_f\cdot\left|l_c\left(\xi-\xi_j\right)\right|/D & 0\leq\left|\xi-\xi_j\right|\leq L_d/l_c \\ 2\left(-\tau_s\cdot\left|l_c\left(\xi-\xi_j\right)\right|+L_d\left(\tau_s-\tau_f\right)\right)/D & L_d/l_c\leq\left|\xi-\xi_j\right|\leq\left(L'_j-L_d\right)/l_c\end{cases} \quad,\text{ for } j=1,2,3,...,N' \quad (25)$$

The stresses in composite with debonding can be obtained again by using the developed superposition method along with eqs. (22-25).

4. Statistical modelling

The catastrophic failure of fiber-reinforced composites is primarily dominated by the failure of fibers (Talrejia, 1995). The fibers typically exhibit variability in strength due to microflaws distributed randomly along the length. The variation of the fiber strength σ_f for length L can be characterized by the probability density function, $F(\sigma_f)$, which is assumed to follow a two-parameter Weibull distribution, that is,

$$F\left(\sigma_f\right) = 1 - \exp\left\{-\left(\frac{L}{L_0}\right)\left(\frac{\sigma_f}{\sigma_0}\right)^{\beta}\right\}, \tag{26}$$

where σ_0 represents the scale parameter, or characteristic quantity of the material. L_0 is the reference length when the characteristic fiber strength σ_0 is measured. The Weibull modulus or shape parameter β controls the scatter of the fiber tensile strength in the distribution, experimentally found to describe a variety of materials. This scatter will become large with decreasing β. The parameters β and σ_0 can be calculated by statistical method

$$\bar{\sigma}_c = E\left(\sigma_c\right) = \sigma_0 \Gamma\left(1 + \frac{1}{\beta}\right), \tag{27}$$

$$S^2 = D\left(\sigma_c\right) = \sigma_0^2 \left\{\Gamma\left(1 + \frac{2}{\beta}\right) - \left[\Gamma\left(1 + \frac{1}{\beta}\right)\right]^2\right\}, \tag{28}$$

where $E(\sigma_c)$ and $D(\sigma_c)$ are the mean and variance of random variable, respectively.

In the approach we use here, we assume the total length of composite specimen to be modeled, $2L_T$, is divided into N_L segments of equal length Δx such that $\Delta x = 2L_T/N_L$. The composite specimen to be modeled contains N_f fibers. For a given failure probability F the strength of the fiber segments with length Δx can be derived from the inversion of eq. (26)

$$\sigma_f = \sigma_0 \left\{\left(\frac{L_0}{\Delta x}\right)\ln\left(\frac{1}{1-F}\right)\right\}^{1/\beta}, \tag{29}$$

in which the failure probability F is a random number taken from the uniform distribution in the range $[0,1]$, then the strength of each fiber segment can be obtained. By introducing the normalized fiber segment length $\hat{L} = \Delta x / D$, eq. (29) can be written as

$$\sigma_f = \sigma^* \left(\frac{1}{\hat{L}}\right)^{1/\beta} \left[\ln\left(\frac{1}{1-F}\right)\right]^{1/\beta}, \tag{30}$$

where σ^* acts as the scale parameter and $\sigma^* = \sigma_0 \cdot (L_0/D)^{1/\beta}$.

The Monte-Carlo simulation technique coupled with the proposed analytical model is executed to simulate the mechanical failure process in fiber-reinforced composites under TMF loads. At the beginning of each simulation, all fiber segments are assumed to be intact. Their status changes from intact to fractured if the fiber stresses satisfy the failure criteria. To employ the overall method described above, the algorithm is follows,

a. Randomly assign each fiber element a strength according to a two-parameter Weibull distribution.
b. From the displacement and the boundary condition, the stress for each segment of fiber, matrix, and interface are then obtained by using the stress analysis.
c. Determine whether the fiber elements will break up or not, whether the matrix and interface around broken fibers will yield or not for each incremental load. If no new damage occurs at this loading step, we calculate the composite stresses. Otherwise, we recalculate the stress field by taking account of the new breakage, and repeat this step until no more damage occurs at the load level.
d. Taking unloading. For the case except of complete elastic unloading, the cyclic plastic strain range along with Coffin-Manson equation is be used to predict the debonding
e. Increase a new loading by a small increment and repeat step (b) and (c) and (d), until the stress-strain curve up to failure for the composite is obtained.

For each loading step at which the applied composite strain is given the overall composite stress can be evaluated by

$$\sigma_{comp} = V_f \tilde{\sigma}_f + (1 - V_f) \tilde{\sigma}_m, \tag{31}$$

where V_f denotes the fiber volume fraction; the symbols $\tilde{\sigma}_f$ and $\tilde{\sigma}_m$ are used for the average fiber and matrix stress, respectively.

The average fiber stress is computed by

$$\tilde{\sigma}_f = \frac{1}{2 N_f L_T} \sum_{k=1}^{N_f} \left[\int_{-L_T / l_c}^{L_T / l_c} \sigma_f(k, \xi) d\xi \right], \tag{32}$$

and the average matrix stress is evaluated by

$$\tilde{\sigma}_m = \frac{1}{2 N_m L_T} \sum_{k=1}^{N_m} \left[\int_{-L_T / l_c}^{L_T / l_c} \sigma_m(k, \xi) d\xi \right], \tag{33}$$

The fatigue failure of the fiber reinforced ductile composites occurs as a result of accumulation of the large amounts of damage-plasticity events under cyclically thermomechanical loading. As much more fiber breaks and the associated local thermoplasticity are accumulated, the composite as a whole will be unable to carry additional load and fail will ensue.

5. Predictions and concluding remarks

For illustrations, the Boron/Al continuous fiber reinforced composite is examined. The properties of the constituents are given below

$$E_f = 400 \text{ GPa}, \quad E_m = 70.2 \text{ GPa}, \quad \alpha_f = 6.3 \ \mu\varepsilon/^\circ C, \quad \alpha_m = 23.9 \ \mu\varepsilon/^\circ C,$$

$$D = 0.14 \text{ mm}, \quad V_f = 0.48, \quad c = -0.65, \quad \gamma_f = 0.42$$

Fig. 4. shows a simulated failure process of the fiber-reinforced composite under in-phase condition. In this figure, 4a-f indicate the damage configurations under some instantaneous stage, respectively. Because of large variability for fiber strength, the fiber element with the lowest strength is firstly broken at the early loading stage, see Fig. 4a. The high stress concentrations are generated in the matrix or the interface due to the fiber breakages, so that plastic yield appears for matrix or interface. As the applied load increases, more fiber breakages occur in the whole specimen, see Fig. 4b-c. Stress redistribution in the composite is caused by debonding after some cycles, shown in Figure 4d. Accumulation of severe damages can be observed until the composite completely fails, in Fig. 4e-f.

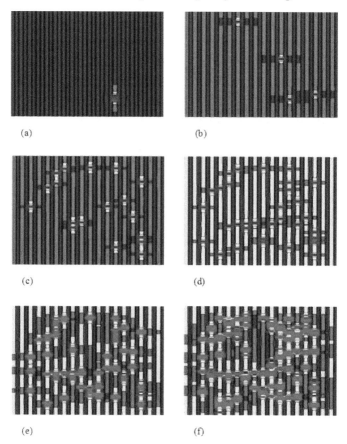

Figure 4. Progressive damage of advanced fiber composites.

Fig. 5. illustrates the cyclic stress/strain response corresponding to the example in Fig. 4. Plastic yield onset in unloading and reloading is represented by the '+' symbols. Due to plastic deformation, the curve deviates from linear behavior beyond the elastic range at unloading and reloading state. It is distinctly seen that the composite strain accumulates with cycles under the constant stress amplitude. This attributes to the microdamage mechanism that a dominant 'critical cluster' of breaks is shaped, which leads to the failure of the rest of fibers.

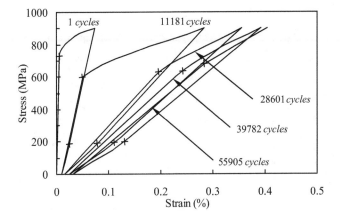

Figure 5. Predicted cyclic nonlinear stress-strain response.

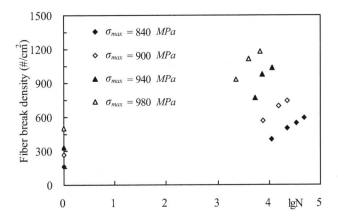

Figure 6. Evolution of fiber breakage with cycles under different load amplitude.

In order to show accumulation of damage-plasticity in the composite, Figs. 6-8. illustrate the evolution of fiber break density, matrix yield density, and the total debonding length with TMF cycles, respectively. It is obvious that the deformation behavior is dependent on the load amplitude: the degree of damage exhibits a more aggravation with cycle times under higher amplitude.

Fig. 9. shows the evolution of fiber cracking in the composites during reloading. Initial breaking takes place at about 750 MPa in the first cycles. The density of broken fiber increases as the number of load cycles increases. It should be emphasized that this is due to the stress redistribution induced by progressive debonding.

Fig. 10. is a plot of fatigue life by logarithm coordinate against maximum stress. The predicted S-N curves for the in-phase and out-of-phase TMF loads with a temperature range from 250 to 350 °C are plotted along with the experimental S-N curves tested by others (Nicholas, 1995; Liu & He, 2001).

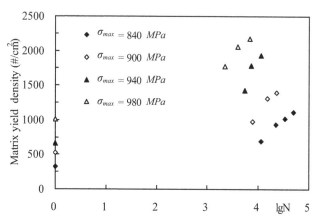

Figure 7. Evolution of matrix tensile yielding with cycles under different load amplitude.

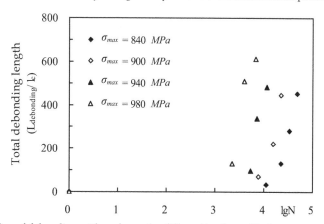

Figure 8. Evolution of debonding with cycles under different load amplitude.

In high stress level, the fatigue life for in-phase TMF conditions is considerably shorter than the life for the out-of-phase TMF conditions. On the other hand, in low stress level, the fatigue life for in-phase conditions is seen to be longer considerably than the life for out-of-phase conditions. This cross-over behavior is due to the difference of the dominated failure mode for this two TMF conditions. A micro-mechanism is presented in Figs. 11-12. Micro failure mechanism associated with the in-phase TMF is characterized by a fiber-dominated fracture with evidence of smaller matrix yielding. Evidence of larger matrix plasticity and less fiber fracture, suggestive of a matrix-dominated failure under the out-of-phase TMF, has been observed in the model simulation.

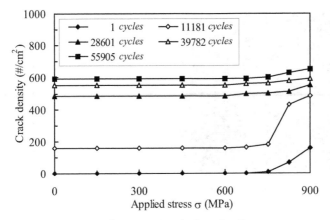

Figure 9. Evolution of fiber cracking in the composites during reloading.

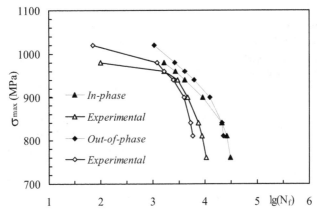

Figure 10. Comparison of model predictions with experimental results for TMF loads.

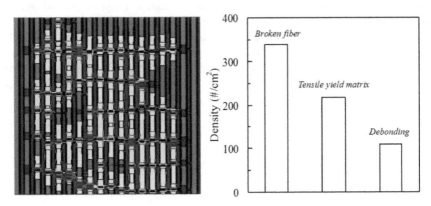

Figure 11. Damage-plasticity configuration when the composite completely fails under in-phase condition.

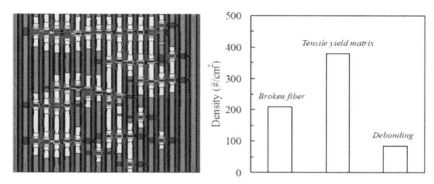

Figure 12. Damage-plasticity configuration when the composite completely fails under out-of-phase condition.

6. Conclusions

Modelling of the mechanical behavior of fibrous composite under progressive damages attracts much attention for a long time. The methodology developed in this paper provides an engineering tool to investigate damage evolution in fiber-reinforced composites, especially under cyclical thermomechanical loading. It has to be point out that we do not intend to make the quantitative comparison of the model with the experimental data. The predicted lifetime of fibrous composite is substantially higher than the measured, especially for the regime of high load level. There are several factors that may cause the quantitative discrepancy between the predicted and measured fatigue life. First, the model ignored the interface normal stress that dominates the initiation of debonding, leading to an estimate of delayed debonding and a longer lifetime of the composite. Second, the interface properties have been assumed to be the same as those of matrix; however, interfaces may contain inherent flaws or defects. This assumption will result in a longer lifetime of debonding growth.

As a final remark, it should be mentioned that the 2D modeling presented here does generalize the governing equations to include interactions with multiple damage events and develop the basic mechanics that are necessary to understanding failure mode of the composite produced by the failure of one or more of its components from a micromechanics perspective. The difference between predicted values and measured ones suggests, however, that consideration of the effect of variations in fiber strengths alone is not sufficient for predicting the variability of composite strength. A more precise calibration of the model, capable of explaining such effects as 3D fiber arrays and fiber/matrix interface sliding on composite strength, is the subject of our further study.

Author details

F. Wang
Southwest University, China

J.Q. Zhang
Shanghai University, China
Shanghai Key Laboratory of Mechanics in Energy Engineering, China

Acknowledgement

The chapter was written with the financial support of the National Science Foundation of China under Grant No.11172159, 11102169, 10772105 and 10372120, and Natural Science Foundation Project of CQ CSTC, 2009BB4290.

7. References

Beyerlein, I.J. & Landis, C.M. (1999). Shear-lag model for failure simulations of unidirectional fiber composites including matrix stiffness. *Mechanics of Materials*, Vol.31, No.5, (April, 1999), pp. 331-350, ISSN 0167-6636.

Beyerlein, I.J. & Phoenix, S.L. (1996). Stress concentrations around multiple fiber breaks in an elastic matrix with local yielding or de-bonding using quadratic influence superposition, *Journal of the Mechanics and Physics of Solids*, Vol.44, No.12, (August, 1996), pp. 1997-2039, ISSN 0022-5096.

Blassiau, S., Thionnet, A. & Bunsell, A.R. (2008). Micromechanisms of load transfer in a unidirectional carbon fiber-reinforced epoxy composite due to fiber failures: part 3. Multiscale reconstruction of composite behavior, *Composite Structures*, Vol.83, No.3, (May, 2007), pp. 312-323, ISSN 0263-8223.

Bussiba, A., Kupiec, M., Ifergane, S., Piat, R. & Böhlke, T. (2008). Damage evolution and fracture events sequence in various composites by acoustic emission technique, *Composites Science and Technology*, Vol.68, No.5, (August, 2007), pp. 1144-1155, ISSN 0266-3538.

Curtin, W.A. (1999). Stochastic damage evolution and failure in fiber-reinforced composites, In: *Advances in Applied Mechanics*, E.V.D. Giessen & T.Y. Wu, (Eds.), pp. 163-253, Academic Press, ISBN 978-012-0020-36-2, San Diego, USA.

González, C., Segurado, J. & LLorca J. (2004). Numerical simulation of elasto-plastic deformation of composites: evolution of stress microfields and implications for homogenization model, *Journal of the Mechanics and Physics of Solids*, Vol.52, No.7, (January, 2004), pp. 1573-1593, ISSN 0022-5096.

Hanan, J.C., Üstündag, E., Beyerlein, I.J., Swift, G.A., Almer, J.D., Lienert, U. & Haeffner, D.R. (2003). Microscale damage evolution and stress redistribution in Ti-SiC fiber composites, *Acta Materialia*, Vol.51, No.14, (April, 2003), pp. 4239-4250, ISSN 1359-6454.

Hanan, J.C., Mahesh, S., Üstündag, E., Beyerlein, I.J., Swift, G.A., Clausen, B., Brown, D.W. & Bourke, M.A.M. (2005). Strain evolution after fiber failure in a single-fiber metal matrix composite under cyclic loading, *Materials Science and Engineering A*, Vol.399, No.(1-2), (February, 2005), pp. 33-42, ISSN 0921-5093.

Kabir, M.R., Lutz, W., Zhu, K. & Schmauder, S. (2006). Fatigue modeling of short fiber reinforced composites with ductile matrix under cyclic loading. *Computational Materials Science*, Vol.36, No.4, (September, 2005), pp. 361-366, ISSN 0927-0256.

Kruch, S., Carrere, N. & Chaboche, J.L. (2006). Fatigue damage analysis of unidirectional metal matrix composites, *International Journal of Fatigue*, Vol.28, No.10, (February, 2006), pp. 1420-1425, ISSN 0142-1123.

Landis, C.M., Beyerlein, I.J. & McMeeking, R.M. (2000). Micromechanical simulation of the failure of fiber reinforced composites. *Journal of the Mechanics and Physics of Solids*, Vol.48, No.3, (January, 2000), pp. 621-648, ISSN 0022-5096.

Liu, S.L. & He, Y.H. (2001). Thermomechanical fatigue behavior of a unidirectional B/Al metal matrix composite, *Proceedings of 13th International Conference on Composite Materials*, pp. 326, ISBN 7-6023-3825-X, Beijing, CHINA, June 25-29, 2001.

Liu, Y.F. (2001). A 3-d micromechanical model of cyclic plasticity in a fiber-reinforced metal matrix composite. *Journal of Materials Science Letters*, Vol.20, No.5, (October, 2000), pp. 415-417, ISSN 0261-8028.

Llorca, J. (2002). Fatigue of particle- and whisker-reinforced metal-matrix composites. *Progress in Materials Science*, Vol.47, No.3, (June, 2000), pp. 283-353, ISSN 0079-6425.

Mall, S. & Schubbe, J.J. (1994). Thermo-mechanical fatigue behavior of a cross ply SCS-6/Ti-15-3 metal matrix composite. *Composites Science and Technology*, Vol.50, No.1, (February, 1993), pp. 49-57, ISSN 0266-3538.

Miracle, D.B. (2005). Metal matrix composites-From science to technological significance. *Composites Science and Technology*, Vol.65, No.(15-16), (May, 2005), pp. 2526-2540, ISSN 0266-3538.

Mishnaevsky, L. & Brøndsted, P. (2008). Three-dimensional numerical modelling of damage initiation in unidirectional fiber-reinforced composites with ductile matrix. *Materials Science and Engineering A*, Vol.498, No.(1-2), (September, 2007), pp. 81-86, ISSN 0921-5093.

Nicholas, T. (1995). An approach to fatigue life modeling in titanium-matrix composites. *Materials Science and Engineering A*, Vol.200, No.(1-2), (December, 2004), pp. 29-37, ISSN 0921-5093.

Talrejia, R. (1995). A conceptual frame work for interpretation of MMC fatigue. *Materials Science and Engineering A*, Vol.200, No.(1-2), (December, 2004), pp. 21-28, ISSN 0921-5093.

Williams, T.O. & Pindera, M.J. (1995). Thermo-mechanical fatigue modeling of advanced metal matrix composites in the presence of microstructural details. *Materials Science and Engineering A*, Vol.200, No.(1-2), (May, 1995), pp. 156-172, ISSN 0921-5093.

Xia, Z., Curtin, W.A. & Peters, P.W.M. (2001). Multiscale modeling of failure in metal matrix composites. *Acta Materialia*, Vol.49, No.2, (January, 2001), pp. 273-287, ISSN 1359-6454.

Xia, Z., Okabe, T. & Curtin, W.A. (2002). Shear-lag versus finite element models for stress transfer in fiber-reinforced composites. *Composites Science and Technology*, Vol.62, No.9, (October, 2001), pp. 1141-1149, ISSN 0266-3538.

Zhang, F., Lisle, T., Curtin, W.A. & Xia, Z. (2009). Multiscale modeling of ductile-fiber-reinforced composites. *Composites Science and Technology*, Vol.69, No.(11-12), (April, 2009), pp. 1887-1895, ISSN 0266-3538.

Zhang, J.Q., Wu, J. & Liu, S.L. (2002). Cyclically thermomechanical plasticity analysis for a broken fiber in ductile matrix composites using shear lag model. *Composites Science and Technology*, Vol.62, No.5, (February, 2002), pp. 641-654, ISSN 0266-3538.

Zhang, J.Q. & Wang, F. (2009). Modeling of progressive failure in ductile matrix composites including local matrix yielding. *Mechanics of Advanced Materials and Structures*, Vol.16, No.7, (March, 2009), pp. 522-535, ISSN 1537-6494.

Zhang, J.Q. & Wang, F. (2010). Modeling of damage evolution and failure in fiber reinforced ductile composites under themomechanical fatigue loading. *International Journal of Damage Mechanics*, Vol.19, No.7, (January, 2010), pp. 851-875, ISSN 1056-7895.

Chapter 9

Molecular Simulations on Interfacial Sliding of Carbon Nanotube Reinforced Alumina Composites

Yuan Li, Sen Liu, Ning Hu, Weifeng Yuan and Bin Gu

Additional information is available at the end of the chapter

1. Introduction

With remarkable physical and mechanical properties [1, 2], carbon nanotube (CNT), either single-walled carbon nanotube (SWCNT) or multi-walled carbon nanotube (MWCNT), has prompted great interest in its usage as one of the most promising reinforcements in various matrices (e.g., polymers, metals and ceramics) [3-11]. However, the dramatic improvement in mechanical properties has not been achieved so far. The reason can be attributed to several critical issues: (1) insufficient length and quality of CNT, (2) poor CNT dispersion and alignment, and (3) weak interface between CNT and matrix. Although great progress has been made to improve the first two issues by developing newly cost-effective CNT synthesis methods and exploring specific fabrication methods of composites (e.g., spark plasma sintering [12], sol-gel process [13]), the proper control of interfacial properties is still a challenge as the inherent characteristics is unclear.

Up to date, large amounts of investigations have been focused on the interfacial properties of polymer-based composites by using direct pull-out experiments with the assistance of advanced instruments (e.g., transmission electronic microscopy (TEM) [14,15], atom force microscope (AFM) [16,17], Raman spectroscopy [18], scanning probe microscope (SPM) [19]), or theoretical analysis based on continuum mechanics (e.g., cohesive zone model [20], Cox's model [21], shear lag model [22,23] and pull-out model [24,25]), or atomic simulations [26-33]. However, in contrast, much less work has been focused on the interfacial properties of alumina-based composites [34-38]. For example, it has been reported that there are three hallmarks of toughening behavior demonstrated in CNT-reinforced alumina composites (CNT/Alumina) as below [34]: crack deflection at the CNT/Alumina interface; crack bridging by CNT, and CNT pull-out on the crack plane, which is consistent with that in conventional micron-scale fiber reinforced composites. Therefore, a fundamental understanding on the interfacial sliding between CNT and alumina matrix (i.e., CNT pull-out from alumina matrix) is important for clarifying the interfacial properties, and therefore the mechanical properties of bulk CNT/Alumina composites.

Current experimental works have reported two common sliding behaviors in CNT/Alumina composites: the pull-out of SWCNT [35] and sword-in-sheath mode [36, 37] of MWCNT (i.e. "pull-out of the broken outer walls of CNT with matrix", or "pull-out of inner walls of CNT with matrix after the breakage of the outer walls" in relativity). Therefore, clarifying the above two distinguished pull-out behaviors is of critical importance for understanding the interfacial properties of CNT-reinforced composites.

In this Chapter, a series of pull-out simulations of either SWCNT or MWCNT from alumina matrix are carried out based on molecular mechanics (MM) to investigate the corresponding interfacial sliding behaviors in CNT/Alumina composites. By systematically evaluating the variation of potential energy increment during the pull-out process, the effects of grain boundary (GB) structures of alumina matrix, nanotube length, nanotube diameter, wall number and capped structure of CNTs are explored for the first time.

2. Computational model

As experimentally identified, CNTs are generally located in the GB of alumina [34-37, 39], which can be schematically illustrated in Fig. 1. Note that the GB structure is generally characterized by a multiplicity index Σ based on the geometrical concept of three-dimensional (3D) coincidence between two crystals named the coincidence site lattice (CSL) model [40], which is defined as the ratio of the crystal lattice sites density to the density of the two grain superimposed lattices.The corresponding computational model by using the commercial software of Materials Studio (Accelrys) can be constructed as follows:

1. Building a hexagonal primitive cell of neutral alumina;
2. Cleaving the required GB planes and joining them together;
3. Inserting a CNT into the GB;
4. Relaxing the constructed model to obtain the equilibrated configuration.

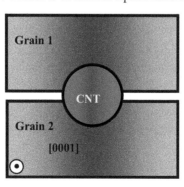

Figure 1. Schematic GB with CNT

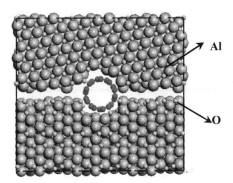

Figure 2. Simulation cell of GB with CNT

As an example, the equilibrated model of Σ7 (14 $\bar{5}$ 0)//(41 $\bar{5}$ 0) [41] GB is shown in Fig. 2, in which the inserted open-ended SWCNT (5,5) has the length of l=5.17nm and diameter of D=0.68nm.

The pull-out process of CNT is schematically given in Fig.3, which is mainly divided into the following two steps:

1. Applying the fixed boundary conditions to the left end of alumina matrix;
2. Pulling out the CNT gradually along its axial (x-axis) direction with a constant displacement increment Δx of 0.2nm.

After each pull-out step, the structure should be relaxed in order to obtain the minimum systematic potential energy E.

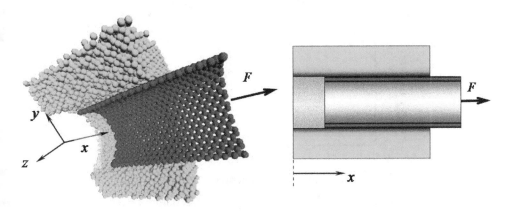

Figure 3. Schematic pull-out process of CNT from alumina matrix
(green balls: atoms in alumina matrix, purple balls: atoms of CNT)

3. Pull-out simulations of open-ended SWCNTs from alumina matrix

3.1. Effect of GB structure

To a large extent, GBs play a significant role on the microstructure formation and properties control of polycrystalline materials. To explore the influence of GB structure on the interfacial siding behavior between CNT and alumina matrix, three representative GB structures with a common rotation axis of [0001] ($\Sigma3(11\bar{2}0)//(11\bar{2}0)$ [42], $\Sigma7(14\bar{5}0)//(41\bar{5}0)$ [41,43,44], and $\Sigma31(47\bar{11}0)$ // $(74\bar{11}0)$ [44]) are modeled. Note that the same fragment of SWCNT(5,5) with the length of l=5.17nm and diameter of D=0.68nm is employed.

The obtained variations of energy increment ΔE between two consecutive pull-out steps are plotted in Fig. 4, where three distinct stages can be clearly seen for each case. In the initial ascent stage I, ΔE increases sharply until the pull-out displacement x reaches up to about 1.0nm. After that ΔE undergoes a long platform stage II followed by the quick descent stage III until the complete pull-out. It is noticeable that both stages I and III have the same range corresponding to the pull-out displacement of approximately a=1.0nm, which is very close to the cut-off distance of vdW interaction (i.e., 0.95nm). This feature of ΔE is similar to that for the pull-out process of CNT from polymer matrix [33] and that for the sliding among nested walls in a MWCNT [45].

Moreover, ΔE in these three curves are almost identical in stages I and III, and have the same average value at stage II although $\Sigma7$ GB results in a slightly higher ΔE. This suggests that the GB structure of alumina matrix has only a limited effect on the energy increment ΔE between two adjacent pull-out steps.

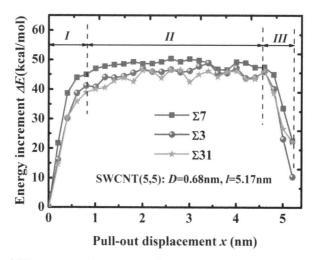

Figure 4. Effect of GB structure on the variation of energy increment during the pull-out process

Here, as discussed in Refs. [33,45], the stable pull-out stage II is focused on, in which the average energy increment in stage II is referred to as ΔE_{II} hereinafter. Obviously, ΔE_{II} is *independent of GB structure of alumina matrix.* Therefore, in the following simulations, the Σ31 GB structure is employed to investigate the effect of nanotube length and diameter on the pull-out process.

3.2. Effect of nanotube length

To investigate the effect of nanotube length on the pull-out process, three SWCNTs (5,5) with different lengths are embedded in the same Σ31 GB of alumina matrix, respectively. The obtained variations of energy increment ΔE between two adjacent pull-out steps are given in Fig.5, in which the same trend is clearly observed for each case as that in Fig. 4. Moreover, the identical ΔE_{II} of three cases indicates its *independence of nanotube length.* Therefore, in the following simulations, CNTs with the same length of 5.17nm are employed.

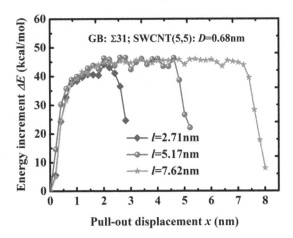

Figure 5. Effect of nanotube length on the variation of energy increment during pull-out process

3.3. Effect of nanotube diameter

Based on the above length-independent behavior, four SWCNTs (i.e., (5,5), (10,10), (15,15), (20,20)) with the same length of 5.17nm but different diameters are embedded into alumina matrix with Σ31GB structure. The corresponding relationship between energy increment ΔE and pull-out displacement x is shown in Fig.6a. Unlike the length-independent behavior, ΔE_{II} *increases linearly with nanotube diameter* as fitted in Fig.6b with the following formula

$$\Delta E_{II} = 52.04 \times D + 9.04 \tag{1}$$

where ΔE_{II} is in kcal/mol, and D is in nm.

This phenomenon can be attributed to the number of atoms in circumferential direction, which increases linearly with nanotube diameter. For a CNT with larger diameter, there will be stronger vdW interactions needed to be overcome for the possible pull-out, which subsequently induces the higher energy increment in the same pull-out displacement.

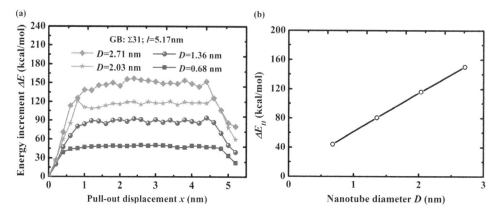

Figure 6. Effect of nanotube diameter on the variation of energy increment during the pull-out process (a) Relationship of energy increment and pull-out displacement; (b) Relationship of ΔE_{II} and nanotube diameter

3.4. Pull-out force and surface energy density

As discussed above, for the pull-out of a SWCNT, the corresponding average energy increment in stage *II*, i.e., ΔE_{II}, is independent of GB structure and nanotube length, but is proportional to nanotube diameter. In view of that the work done by the pull-out force is equal to the energy increment in each pull-out step by neglecting some other minor energy dissipations, the pull-out force can be approximately calculated as

$$F_{II} = \frac{\Delta E_{II}}{\Delta x} \qquad (2)$$

On this sense, we can conclude that *the pull-out force of CNT from alumina matrix related to energy increment is also independent of GB orientation and nanotube length, but is proportional to nanotube diameter*. From Eqs. (1) and (2), the corresponding empirical formula to predict the pull-out force is proposed as

$$F_{II} = 1.81 \times D + 0.31 \qquad (3)$$

where F_{II} is in the unit of nN, and D of nm.

It should be noted that two new surface regions are generated at two ends of CNT after each pull-out step (i.e., the inner surface of the matrix at the left side of CNT, and the outer surface of CNT on the right side). Therefore, the corresponding surface energy should be equal to the energy increment. Therefore, the surface energy density can be calculated as

$$\gamma_{II} = \frac{\Delta E_{II}}{2\pi D \Delta x} = \frac{F_{II}}{2\pi D} \qquad (4)$$

Initially, this value is dependent on the diameter of SWCNT. However, as nanotube diameter increases, it will decrease gradually and then saturate to a constant. The converged value of surface energy density is approximately 0.3N/m. Note that this surface energy density is newly reported for the interface of SWCNT and alumina matrix, although there have been some reports about that for the interface of SWCNT and polymer matrix with the value of 0.09~0.12N/m [26, 30] or for sliding interface among nested walls in a MWCNT with the value of 0.14N/m [45]. It can be found that the surface interface density in CNT/Alumina composites is much higher than those of CNT/Polymer composites or CNT walls, implying its stronger interface.

3.5. Interfacial shear stress

Based on the above discussion, the corresponding interfacial shear stress is analyzed in the following.

The pull-out force is equilibrated with the axial component of vdW forces which induces the interfacial shear stress. Conventionally, if we employ the common assumption of constant interfacial shear stress with uniform distribution along the whole embedded region of CNT, the pull-out force F_{II} will vary with the embedded length of CNT, which is obviously in contradiction to the above length-independent behavior of average energy increment ΔE_{II} in stage II. For the extreme case of a CNT with an infinite length, the interfacial shear stress tends to be zero, which is physically unreasonable. This indicates that the conventional assumption of interfacial shear stress is improper for the perfect interface of CNT/Alumina composites with only consideration of vdW interactions.

For this problem, the interfacial shear stress should be analyzed according to the different stages in the variation of energy increment ΔE. In stage I, the interfacial shear stress exists within a region of the length a=1.0nm at each end of CNT as described in Ref. [45] since the length of CNT in the model is equal to that of alumina matrix. In stage II, the situation may be different since the left end of CNT is deeply embedded into the alumina matrix with the pull-out displacement x much larger than a=1.0nm. To address the interfacial shear stress in this stage II, a simple simulation is performed here.

As shown in Fig. 7a, a SWCNT(5,5) with only a half repeat unit is completely embedded in the middle position of alumina matrix. Then this SWCNT fragment is pulled out gradually with a constant increment of Δx=0.2nm to obtain the variation of systematic energy increment ΔE_{ω}, and the corresponding pull-out force F_{ω}. As this SWCNT fragment is very short, the obtained pull-out force F_{ω}, which is equilibrated by the shear force induced by the interfacial shear stress, can be used to characterize the distribution of interfacial shear stress. The obtained distribution of pull-out force F_{ω} at various pull-out steps is shown in Fig.7b. At the initial stage of the pull-out, the pull-out force keeps value at zero. When the CNT unit cell moves into the range of a=1.0nm measured from the right end of alumina matrix, the pull-out force increases sharply. It reaches the maximum when the CNT unit cell is just located on the right end of the matrix. As the CNT unit cell is further pulled out, it decreases gradually to zero. In virtue of the above results, as shown in Fig. 7c, the interfacial shear stress is solely distributed within the region of $2a$ centered by the right end of matrix in stage *II*. The pull-out force during the pull-out process is further averaged within the range of 2*a*, i.e., F_ω^* =0.09nN in Fig. 7b.

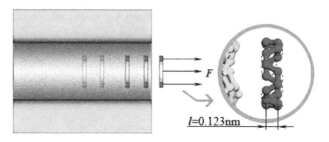

(a) Model for the pull-out of a simple CNT unit cell

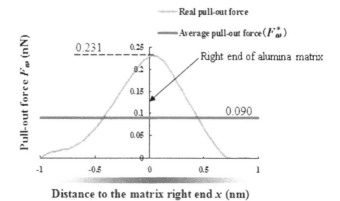

(b) Variation of pull-out force F_ω and average pull-out force F_ω^*

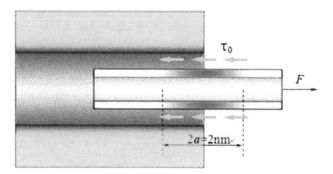

(c) Distribution of average interfacial shear stress

Figure 7. Analysis of interfacial shear stress

By assuming that the interfacial shear stress is uniform within the above defined region for simplicity, the average of interfacial shear stress τ_0 in stage II can be defined from the pre-defined average pull-out force(i.e., F_ω^*) as

$$\tau_0 = \frac{F_\omega^*}{2\pi Da} \tag{5}$$

Obviously, τ_0 is dependent on the diameter D of SWCNT. However, it tends to be a constant as nanotube diameter increases gradually. The obtained converged interfacial shear stress τ_0 from various unit cells of SWCNT with different diameters is 303 MPa. Note that it only exists within the range of $2a$ centered by the right end of the matrix.

4. Pull-out simulations of open-ended MWCNTs from alumina matrix

Usually, there are two typical sliding behavior for MWCNT-reinforced composites: one is the complete pull-out of MWCNT, while the other is the so-called sword-in-sheath mode, e.g., in which the broken outer walls are pulled out (i.e., sheath) leaving the intact inner walls (i.e., sword) in the matrix. Therefore, based on the above information, two simple typical cases are firstly investigated: Case 1: pull-out of the whole MWCNT (e.g., Fig. 8a); Case 2: pull-out of only the outermost wall of MWCNT (e.g., Fig. 8b).

In view of the extremely high computational cost, several double-walled carbon nanotubes (DWCNTs) with wall number $n=2$ and triple-walled carbon nanotubes (TWCNTs) with wall number $n=3$ are discussed in the present simulation. The obtained average energy increment ΔE_{II} in stage II related to the pull-out force is also found to be proportional to the diameter of the outermost wall of MWCNT D_o, which can be fitted as

Case 1

$$\begin{cases} \Delta E_{II} = 57.54 \times D_o + 4.36, \ F_{II} = 2.00 \times D_o + 0.15 \ (n=2) & \text{(a)} \\ \Delta E_{II} = 58.26 \times D_o + 6.50, \ F_{II} = 2.03 \times D_o + 0.23 \ (n=3) & \text{(b)} \end{cases} \tag{6}$$

Case 2

$$\begin{cases} \Delta E_{II} = 93.61 \times D_o + 10.17, \ F_{II} = 3.26 \times D_o + 0.35 \ (n=2) & \text{(a)} \\ \Delta E_{II} = 96.60 \times D_o + 10.50, \ F_{II} = 3.33 \times D_o + 0.37 \ (n=3) & \text{(b)} \end{cases} \tag{7}$$

For Case 1, the relationship of ΔE_{II} and nanotube diameter for the complete pull-out of SWCNT (Eq. 3), DWCNT (Eq. 6a), and TWCNT (Eq.7a) are plotted in Fig. 9, which indicates the effect of wall number from some aspect. The slope for DWCNT is about 9.56% higher than that for SWCNT, which highlights the contribution of the first adjacent inner wall to ΔE_{II}. However, the slope of TWCNT is only about 1.24% higher than that for DWCNT, which implies that the contribution of the second inner wall is gradually weakened as the distance from the sliding interface increases. Therefore, it can be concluded that the pull-out of MWCNT from alumina matrix is mostly affected by its two adjacent walls from the sliding interface, which indicates that for the whole pull-out of any

MWCNT with more walls over 3, ΔE_{II} can be approximately assumed to be equal to that of TWCNT (i.e., Eq. 6b).

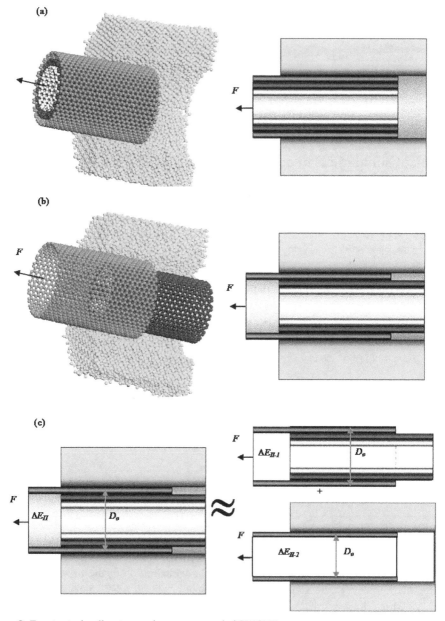

Figure 8. Two typical pull-out cases for an open-ended TWCNT
(a) Case 1: pull-out of the whole MWCNT; (b) Case 2: pull-out of the outmost wall of MWCNT;
c) Decomposition of Case 2 into two independent sub-problems

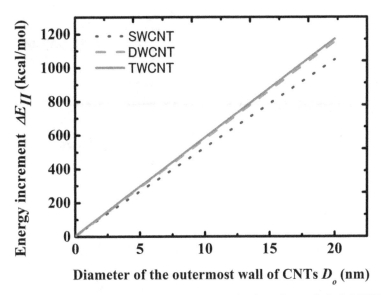

Figure 9. Effect of wall number on the energy increment for the pull-out of whole MWCNT

For Case 2, taking TWCNT(5,5)/(10,10)/(15,15) as an example, it is surprising to find that the corresponding ΔE_{II} is approximately equal to the sum of those of two sub-problems, i.e., the pull-out of a SWCNT(15,15) from alumina matrix, and the pull-out of the outermost wall in the TWCNT. Therefore, the corresponding ΔE_{II} for the pull-out of any TWCNT from alumina matrix (i.e., Eq. 7b) can be approximately decomposed into the following two items as given in Fig. 8c: $\Delta E_{II\text{-}1}$ for the pull-out of the outermost wall of TWCNT against the other two inner walls (i.e., Eq. 5 in Ref. [45]), and $\Delta E_{II\text{-}2}$ for the pull-out of a SWCNT from alumina matrix (i.e., Eq. 3) whose diameter is equal to the outermost wall of the TWCNT.

It should be noted that for the real sword-in-sheath fracture mode, there are more than 3 walls pulled out. For example, as shown in Fig. 10a, several purple outer walls of a MWCNT are pulled out leaving the yellow inner walls within the matrix. Here, there are two sliding interfaces: one is between CNT and matrix, the other is between outer walls and inner wall. According to the above discussion, it can also be thought of as the superimposition of the following two sub-problems in Fig. 10b: one is the pull-out of the TWCNT which is composed of the outer three walls (i.e., Eq. 6b), and the other is the pull-out of outer three walls in a MWCNT with five walls (i.e., Eq. 6 in Ref. [45]). It indicates that the corresponding ΔE_{II} and pull-out force F_{II} can be calculated as

$$\Delta E_{II} = 58.26 \times D_0 + 37.56 \times D_c - 4.00 \tag{8}$$

$$F_{II} = 2.04 \times D_0 + 1.31 \times D_c - 0.14 \tag{9}$$

Here, D_0 is the diameter of the outermost wall of MWCNT, and D_c is the diameter of the green critical wall in Fig. 10 (i.e., the immediate outer wall at the sliding surface between outer walls and inner walls).

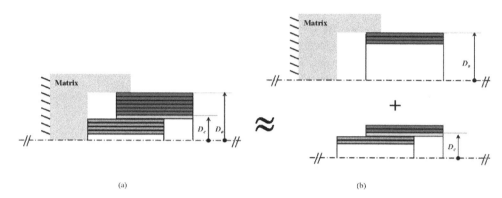

(a)   (b)

Figure 10. Real case of sword-in-sheath mode

5. Pull-out simulations of a capped MWCNT from alumina matrix

It is noted that open-ended CNTs are employed in the above simulations. On the other hand, it has been reported that CNT cap makes great effect on its field emission properties [46], load transferring ability among nested walls of MWCNT [45, 47, 48]. However, to our best knowledge, there is no any detailed report on the effect of CNT caps on the interfacial properties of CNT-reinforced composites. Therefore, the pull-out of capped MWCNTs from alumina matrix in a sword-in-sheath mode is discussed here.

The schematic model is given in Fig. 11. By using the principal of superimposition, this pull-out process can be decomposed into the following three parts: pull-out of outer walls against matrix (i.e., part I); pull-out of inner walls against outer walls, which can be further decomposed into the open-ended part of inner/outer walls (i.e., part II) and the capped part of inner/outer walls (i.e., part III) as each wall in a MWCNT is composed of open-ended part and capped part.

Generally, the number of broken outer walls and intact inner walls are more than 3. Therefore the corresponding pull-out forces for the above three parts are analyzed as below.

i.   Pull-out of outer walls against matrix (i.e., part I in Fig. 11): According to Eq. 2, the corresponding pull-out force F_1 can be predicted by using Eq. 6b, i.e.,

$$F_1 = 2.03 \times D_o + 0.23 \tag{10}$$

ii. Pull-out of open-ended part of inner walls against outer walls (i.e., part II in Fig. 11): According to Eq. 6 in Ref. [45], the corresponding pull-out force F_2 can be predicted as

$$F_2 = 1.31 \times D_c - 0.37 \tag{11}$$

iii. Pull-out of capped part between inner walls and outer walls (i.e., part III in Fig. 11): This part can be transferred as the interfacial sliding among nested walls in a capped MWCNT.

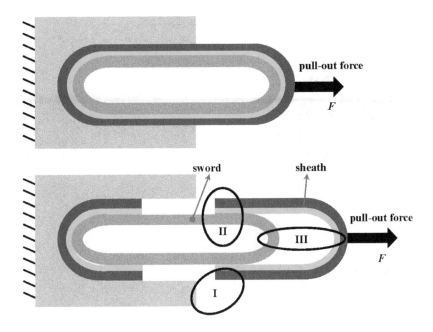

I: outer walls/matrix; II: open-ended part of inner/outer walls; III: capped part of inner/outer walls

Figure 11. Schematic model for the pull-out of a capped MWCNT from alumina matrix

As illustrated in Fig. 12a, after fixing the atoms of the outer cap, the inner wall is pulled out along its axial direction by applying a constant displacement increment of Δx_2=0.01nm on the atoms of the right end of inner wall. Note that the present displacement increment Δx_2 is smaller than the above Δx of 0.2nm, which is used for making the effect of CNT caps on energy increment clearly. After each pull-out step, the structure is relaxed to obtain the minimum potential energy E. As discussed in Ref. [45], the pull-out force of an open-ended CNT is only proportional to nanotube diameter, and independent of nanotube length. For this reason, five DWCNTs with different diameters but same length are built up to investigate the effect of CNT cap. The calculated energy increments between two consecutive pull-out steps of three DWCNTs are shown in Fig. 13a, where D_c is the diameter of critical wall (i.e., the outer wall of DWCNT). It can be seen that for each DWCNT the energy increment ΔE increases rapidly up to a peak value at a specified displacement, and then decreases. The same feature is also observed in the simulations of two other DWCNTs with larger diameters, i.e., (54,54)/(59,59) with D_c=8.0nm and (83,83)/(88,88) with D_c=11.93nm. The maximum energy increment (i.e., ΔE_{max}) for the five DWCNTs is shown in Fig. 13b. The relationship between ΔE_{max} and D_c can be perfectly fitted into a quadratic function of

$$\Delta E_{max-DWCNT} = 2.09 \times D_c^2 - 2.15 \times D_c + 0.94 \tag{12}$$

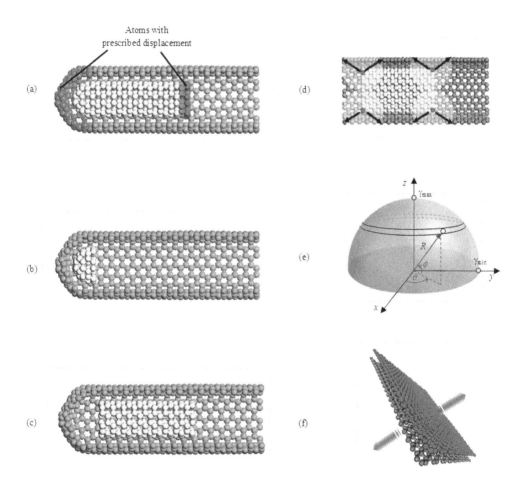

Figure 12. Interfacial sliding in a capped DWCNT
(a) Schematic model of a capped DWCNT; (b) Pull-out of the capped part;
(c) Pull-out of the open-ended part;
(d) Force state of open-ended part of CNT; (e) Estimation of energy variation of a cap;
(f) Pull-out of graphite sheets

To understand this potential energy increment in detail, we further divided the inner wall (Fig. 12a) into two parts, i.e., the capped (Fig. 12b) and the open-ended part (Fig. 12c). The corresponding pull-out forces for these two parts are F_3^1 and F_3^2, which means $F_3 = F_3^1 + F_3^2$. The pull-out of open-ended part (Fig. 12c) does not cause any change of the potential energy, i.e., $F_3^2 = 0$. It means that the contribution of capped part, i.e., F_3^1 dominates the total pull-out force F_3. The reason can be explained using Fig. 12d-12e.

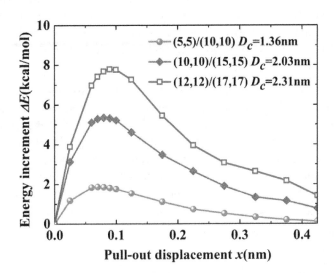

(a) Energy increment versus pull-out displacement for model of Fig.12a

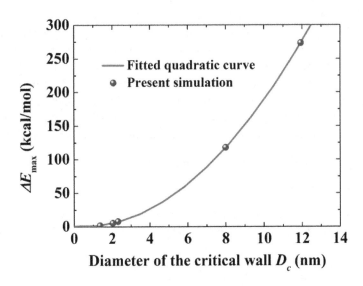

(b) Relationship between maximum energy increment and nanotube diameter

Figure 13. Variation of energy increment for the interfacial sliding in a capped DWCNT

First, in Fig. 12d, if the length of the outer wall of DWCNT is long enough, the carbon atoms of the inner wall are always in force equilibrium. For example, in Fig. 12d, the atoms in red are balanced by the symmetrical horizontal forces form the atoms of the outer wall, which are within the cut-off distance of Lennard-Jones potential [49,50]. During a pull-out process, the relative motion of the atoms between the inner wall and outer wall creates repetitive breaking and reforming of the vdW interactions and no resultant resistance force can be generated on the inner wall, i.e., $F_3{}^2=0$.

The quadratic form of the energy increment in Eq. 12 due to the capped effect is associated with the surface energy density. Considering a cap model shown in Fig. 12e, the bottom edge is just located on the boundary between the capped and open-ended part. If we use γ_{max} and γ_{min} to represent the maximum and minimum surface energy density (i.e., potential energy variation per unit area) under a specified separation displacement, γ_{max} is at the top of the cap while γ_{min} appears at the bottom of the cap. Then, the surface energy density is assumed to vary from the top to the bottom of the cap in the function of $\gamma(\phi) = \gamma_{max} \cos(90° - \phi) = \gamma_{max} \sin\phi$., which implies that $\gamma_{min}=0$. This is reasonable as $F_3{}^2=0$. Then the total surface energy variation of the cap can be calculated as

$$U_{cap} = \int_0^{2\pi} (\int_0^{\frac{\pi}{2}} \gamma(\phi) R\cos\phi \times Rd\varphi)d\theta = \frac{\pi D_c^2}{4}\gamma_{max} \tag{13}$$

From Eq. 13, regardless of the function of $\gamma(\phi)$, the surface energy is always proportional to πD_c^2. As a result, the energy increment induced by the pull-out of the cap can be described by a quadratic function of D_c, which is consistent with Fig. 13b and Eq. 12. Approximately, the γ_{max} at the small top flat area of the cap during the pull-out process can be predicted in the same way by simulating the separation of two flat graphite sheets in Fig. 12f. It is confirmed by the displacement-energy increments curves obtained from the simulation of two graphite sheets which is quite similar with those in Fig. 13a. The corresponding $\gamma_{max-cap}$ is around 0.03N/m under 0.01 nm separation displacement in the normal direction of two graphite sheets. Substituting this value into Eq. 13 leads to the total surface energy change as: $U_{cap} = \frac{\pi D_c^2}{4}\gamma_{max-cap} = 2.76D_c^2$, which is approximately equivalent to Eq. 12. Therefore, it indicates the quadratic form of Eq. 12 is appropriate from the other aspect.

After validating the effectiveness of Eq. 12, the corresponding maximum pull-out force can be simply evaluated by equaling the work done by the pull-out force to the $\Delta E_{max-DWCNT}$ with the formula of

$$F_{3-DWCNT} = 1.45 \times D_c^2 - 1.49D_c + 0.65 \tag{14}$$

It should be noted that the above analysis is for a capped DWCNT. For the case of MWCNT, a simplified model in Fig. 14 is developed, as only the immediate two outer and inner walls from the sliding interface can affect the corresponding pull-out interface [45]. The evaluated pull-out force is found to be approximately 29% higher than that for DWCNT due to the contribution of the immediate two outer and inner walls, which means

$$F_{3-\text{MWCNT}} = 1.87 \times D_c^2 - 1.92D_c + 0.84 \tag{15}$$

Therefore, for the pull-out of a capped MWCNT from alumina matrix in a sword-in-sheath mode, the corresponding pull-out force can be assumed to be the sum of those for the above three parts (i.e., Eq. 10 for part I, Eq. 11 for part II, Eq. 15 for part III):

$$F = F_1 + F_2 + F_{3-\text{MWCNT}} = 1.87 \times D_c^2 + 2.03D_o - 0.61 \times D_c + 0.7 \tag{16}$$

For the pull-out of a MWCNT numbered as sample 14 in Ref. [37] which has the outermost wall with diameter of D_o=94nm and the critical wall at the sliding interface with diameter about D_c=90nm, the calculated pull-out force using Eq. 16 is 15.28µN, which is in the same scale of experimental value of 19.7 µN [37].

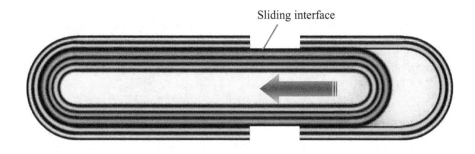

Figure 14. Schematic model for interfacial sliding in a capped MWCNT in sword-in-sheath mode

6. Conclusions

We systematically investigate the pull-out process of open-ended and capped CNTs from alumina matrix using MM simulations, aiming at clarifying the interfacial sliding behavior in CNT/Alumina composites. The effects of grain boundary structure of alumina matrix, nanotube length, nanotube diameter, wall number and capped structure of CNTs are explored systematically.

A set of universal formulae with the newly obtained surface energy density is proposed to approximately predict the pull-out force from nanotube diameter. The philosophy behind these simple empirical formulae is that the pull-out force is only proportional to nanotube diameter, and independent of nanotube length and GB structure of alumina matrix. The detailed interfacial shear stress is studied in this work, which indicates that the conventional definition of the interfacial shear strength is inappropriate in CNT/Alumina composites. Moreover, there are at most two adjacent walls at each side of the sliding interface which will affect this interfacial sliding in CNT/Alumina composites. Furthermore, it also indicates that CNT caps play a very important role in the pull-out process. These findings will be helpful for clarifying the toughening mechanism for mechanical properties of bulk CNT/Alumina composites and providing useful insight into the design of ideal materials.

Author details

Yuan Li

Department of Nanomechanics, Tohoku University, Aramaki-Aza-Aoba, Aoba-ku, Sendai, Japan

Sen Liu

Department of Mechanical Engineering, Chiba University, Yayoi-cho, Inage-ku, Chiba, Japan

Ning Hu

Corresponding Author

Department of Mechanical Engineering, Chiba University, Yayoi-cho, Inage-ku, Chiba, Japan

Weifeng Yuan and Bin Gu

School of Manufacturing Science and Engineering, Southwest University of Science and Technology, Mianyang, P.R.China

Acknowledgement

The authors are grateful to be partly supported by Tohoku Leading Women's Jump Up Project for 2013 (J1 10002158) and Grand-in-Aids for Scientific Research (No. 19360045 and No. 22360044) from the Ministry of Education, Culture, Sports, Science and Technology (MEXT) of Japan. The authors acknowledge Prof. C.B. Fan (Beijing Institute of Technology, China) for kindly providing the computational resources.

7. References

[1] Dresselhaus MS, Dresselhaus G, Avouris P. (2001) Carbon Nanotubes: Synthesis, Structure, Properties, and Applications. Springer-Verlag Berlin Heidelberg.

[2] Dresselhaus MS, Dresselhaus G, Charlier JC, Hernandez E. (2004) Electronic, Thermal and Mechanical Properties of Carbon Nanotubes. Philosophical Transactions of the Royal Society A. 362:2065-2098.

[3] Thostenson ET, Ren ZF, Chou TW (2001) Advances in the Science and Technology of Carbon Nanotubes and their Composites: A Review. Composites Science and Technology. 61(13): 1899-1912.

[4] Breuer O, Sundararaj U. (2004). Big Returns from Small Fibers: A Review of polymer/carbon nanotube composites. Polymer Composites. 25(6): 630-645.

[5] Coleman JN, Khan U, Blau WJ, Gunko YK. (2006) Small but Strong: A Review of the Mechanical Properties of Carbon Nanotube-Polymer Composites. Carbon. 44:1624-1652.

[6] Moniruzzaman M, Winey KI. (2006) Polymer Nanocomposites Containing Carbon Nanotubes. Macromolecules. 39(16): 5194-5205.

[7] Coleman JN, Khan U, Gun'ko YK. (2006) Mechanical Reinforcement of Polymers Using Carbon Nanotubes. Advanced Materials. 18(6): 689-706.

[8] Peigney A, Laurent C, Flahaut E, et al. (2000) Carbon Nanotubes in Novel Ceramic Matrix Nanocomposites. Ceramics International. 26(6): 677-683.

[9] Curtin WA, Sheldon BW. (2004) CNT-reinforced Ceramics and Metals. Materials Today. 7(11):44-49.

[10] Samal SS, Bal S. (2008) Carbon Nanotube Reinforced Ceramic Matrix Composites- a Review. Journal of Minerals & Materials Characterization & Engineering. 7(4): 355-370.

[11] Cho J, Boccaccini AR, Shaffer M. (2009) Ceramic Matrix Composites Containing Carbon Nanotubes. Journal of Materials Science. 44(8): 1934-1951.

[12] Zhan GD, Kuntz JD, Wan J, Mukherjee Ak. (2003) Single-wall Carbon Nanotube as Attractive Toughening Agents in Alumina-based Nanocomposites. Nature Materials. 2:38-42.

[13] Chan BM, Cha SI, Kim KT, Lee KH & Hong SH. (2005) Fabrication of Carbon Nanotube Reinforced Alumina Matrix Nanocomposite by Sol–gel Process. Materials Science and Engineering A. 395: 124–128.

[14] Qian D, Dickey EC, Andrews R, Rantell T. (2000) Load Transfer and Deformation Mechanisms in Carbon Nanotube-Polystyrene Composites. Applied Physics Letters. 76(20):2868-2870.

[15] Deng F. (2008) Investigation of the Interfacial Bonding and Deformation Mechanism of the Nano Composites Containing Carbon Nanotubes. Tokyo University, PhD Dissertation.

[16] Barber AH, Cohen SR, Kenig S, Wagner HD. (2003) Measurement of Carbon Nanotube-Polymer Interfacial Strength. Applied Physics Letters. 82(23): 4140-4142.

[17] Barber AH, Cohen SR, Kenig S, Wagner HD. (2004) Interfacial Fracture Energy Measurements for Multi-walled Carbon Nanotubes Pulled from a Polymer Matrix. Composite Science and Technology. 64:2283-2289.

[18] Schadler LS, Giannaris SC, Ajayan PM. (1998) Load Transfer in Carbon Nanotube Epoxy Composites. Applied Physics Letters. 73(26):3842-3844.

[19] Cooper CA, Cohen SR, Barber AH, Wagner HD. (2002) Detachment of Nanotubes from a Polymer Matrix. Applied Physics Letters. 81(20):3873-3875.

[20] Jiang LY, Huang Y, Jiang H, Ravichandran G, Gao H, Hwang KC, et al. (2006) A Cohesive Law for Carbon Nanotube/Polymer Interfaces based on the van der Waals Force. Journal of the Mechanics and Physics of Solids. 54: 2436- 2452.

[21] Xiao KQ, Zhang LC. (2004) The Stress Transfer Efficiency of a Single-walled Carbon Nanotube in Epoxy Matrix. Journal of Materials Science. 39:4481-4486.

[22] Gao XL, Li K. (2005) A Shear-lag Model for Carbon Nanotube-Reinforced Polymer Composites. International Journal of Solids and Structures. 42: 1649-1667.

[23] Tsai J, Lu T. (2009) Investigating the Load Transfer Efficiency in Carbon Nanotubes Reinforced Nanocomposites. Composite Structures. 90:172-179.

[24] Lau K. (2003) Interfacial Bonding Characteristics of Nanotube/Polymer Composites. Chemical Physics Letters. 370:399-405.

[25] Natsuki T, Wang F, Ni QQ, Endo M. (2007) Interfacial Stress Transfer of Fiber Pullout for Carbon Nanotubes with a Composite Coating. Journal of Materials Science. 42:4191-4196.

[26] Lordi V, Yao N. (2000) Molecular Mechanics of Binding in Carbon-Nanotube-Polymer Composites. Journal of Materials Research. 15(12):2770-2779.

[27] Liao K, Li S. (2001) Interfacial Characteristics of a Carbon Nanotube-Polystyrene Composite System. Applied Physics Letters. 79(25):4225-4227.

[28] Frankland SJV, Caglar A, Brenner DW, Griebel M. (2002) Molecular Simulation of the Influence of Chemical Cross-links on the Shear Strength of Carbon Nanotube- Polymer Interfaces. Journal of Physical Chemistry B. 106:3046-3048.

[29] Gou J, Minaie B, Wang B, Liang Z, Zhang C. (2004) Computational and Experimental Study of Interfacial Bonding of Single-walled Nanotube Reinforced Composites. Computational Materials Science. 31:225-236.

[30] Zheng Q, Xia D, Xue Q, Yan K, Gao X, Li Q. (2009) Computational Analysis of Effect of Modification on the Interfacial Characteristics of a Carbon Nanotube- Polyethylene Composites System. Applied Surface Science. 255: 3524-3543.

[31] Al-Ostaz A, Pal G, Mantena PR, Cheng A. (2008) Molecular Dynamics Simulation of SWCNT-Polymer Nanocomposite and its Constituents. Journal of Materials Science. 43:164-173.

[32] Chowdhury SC, Okabe T. (2007) Computer Simulation of Carbon Nanotube Pull-out from Polymer by the Molecular Dynamics Method. Composites Part A. 38:747-754.

[33] Li Y, Liu Y, Peng X, Yan C, Liu S, Hu N. (2011) Pull-out Simulations on Interfacial Properties of Carbon Nanotube-reinforced Polymer Nanocomposites. Computational Material Science. 50: 1854-1860.

[34] Xia Z, Riester L, Curtin WA, et al. (2004) Direct Observation of Toughening Mechanisms in Carbon Nanotube Ceramic Matrix Composites. Acta Materialia. 52(4):931-944.

[35] Fan, JP, Zhuang DM, Zhao DQ, et al. (2006) Toughening and Reinforcing Alumina Matrix Composite with Single-wall Carbon Nanotubes. Applied Physics Letters B. 89:121910(3).

[36] Yamamoto G, Omori M, Hashida T, Kimura H. (2008) A Novel Structure for Carbon Nanotube Reinforced Alumina Composites with Improved Mechanical Properties. Nanotechnology. 19:315708(7).

[37] Yamamoto G, Shirasu K, Hashida T, et al. (2011) Nanotube Fracture during the Failure of Carbon Nanotube/Alumina Composites. Carbon. 49: 3709-3716.

[38] Li L, Xia Z, Curtin W, Yang Y. (2009) Molecular Dynamics Simulations of Interfacial Sliding in Carbon-Nanotube/Diamond Nanocomposites. Journal of the American Ceramic Society. 92(10): 2331-2336.

[39] Vasiliev AL, Poyato R, and Padture NP. (2007) Single-wall Carbon Nanotubes at Ceramic Grain Boundaries", Scripta Materialia. 56(6): 461-463.

[40] Bollmann W. (1970) Crystal Defects and Crystalline Interfaces. Springer, Berlin.

[41] Matsunaga K, Nishimura H, Hanyu S, et al. (2005) HRTEM Study on Grain Boundary Atomic Structures Related to the Sliding Behavior in Alumina Bicrystals. Applied Surface Science. 241: 75-79.

[42] Ikuhara Y. (2001) Grain Boundary and Interface Structure in Ceramics. Journal of the Ceramic Society of Japan. 109(7): S110-S120.

[43] Nishimura H, Matsunaga K, Saito T, et al. (2003) Atomic Structures and Energies of $\Sigma 7$ Symmetrical Tilt Grain Boundaries in Alumina Bicrystals. Journal of the American Ceramic Society. 86(4): 574-580.

[44] Nishimura H, Matsunaga K, Saito T, et al. (2003) Grain Boundary Structures and High Temperature Deformations in Alumina Bicrystals. Journal of the Ceramic Society of Japan. 111(9): 688-691.

[45] Li Y, Hu N, Yamamoto G, Wang Z, et al. (2010) Molecular Mechanics Simulation of the Sliding Behavior between Nested Walls in a Multi-walled Carbon Nanotube. Carbon. 48:2934-2940.

[46] Wang MS, Wang JY, Peng LM. (2006) Engineering the Cap Structure of Individual Carbon Nanotubes and Corresponding Electron Field Emission Characteristics. Applied Physics Letters. 88(24): 243108(1-3).

[47] Shen GA, Namilae S, Chandra N. (2006) Load Transfer Issues in the Tensile and Compressive Behavior of Multi-walled Carbon Nanotubes. Materials Science and Engineering A. 429(1-2):66-73.

[48] Xia Z, Curtin WA. (2004) Pullout Forces and Friction in Multiwall Carbon Nanotubes. Physical Review B. 69:2333408(1-4).

[49] Hu N, Fukunaga H, Lu C, Kameyama M, Yan B. (2005) Prediction of Elastic Properties of Carbon Nanotube Reinforced Composites. Proceedings of the Royal Society A. 461:1685-1710.

[50] Hu N, Nunoya K, Pan D, Okabe T, Fukunaga H. (2007) Prediction of Buckling Characteristics of Carbon Nanotubes. International Journal of Solids and Structures. 44:6565-6550.

Design, Processing, and Manufacturing Technologies

Chapter 10

Netcentric Virtual Laboratories for Composite Materials

E. Dado, E.A.B. Koenders and D.B.F. Carvalho

Additional information is available at the end of the chapter

1. Introduction

Physical laboratory-based experiments and testing has been a way to develop fundamental research and learning knowledge for many areas of (civil) engineering education, science and practice. In the context of education, it has particularly enriched engineering education by helping students to understand fundamental principles and by supporting them to understand the link between the theoretical equations of their text books and real world applications. In the context of science, physical laboratory experiments have been used to scrutinize particular phenomenon in a real-life setting or to verify and validate scientific computational models over a longer period of time. In both the educational and research context, conducting physical laboratory experiments are generally governed by complex and expensive lab-infrastructures and require significant allocation of resources from the educational and research institutes. Besides, results most frequently have a limited range of exposure and are available for a relatively small audience (i.e. high costs versus relatively low benefits) [1]. In the context of practice, physical tests are often performed to validate performances of products. With the increasing regulations from national governments and European Union (EU) concerning quality, safety and environmental properties of products, the number of physical tests performed in laboratories of certificated (research) institutes have increased recently. For example, the European labels of conformity, known as CE markings, are a guarantee of quality and safety for products produced and sold in the EU. According to the CE conformity standards, products will have undergone a series of performance tests before they can be sold on the EU market. However, these performance tests entail additional costs which may result in financial difficulties stemming from these additional costs and in the long run result in competitive disadvantages for Small and Medium Enterprises (SMEs) in Europe [2].

To improve this situation, initiatives have been launched where research and development (R&D) projects could be conducted in so-called 'virtual laboratories'. In civil engineering, the National Institute of Standards and Technologies (NIST) in the United States were the first who set the standard for R&D projects conducted in a virtual laboratory for composite materials [3]. From this initiative and the promises of emerging information and communication technologies a whole new realm of possibilities for developing virtual laboratories has become available. Based on these observations and developments, the authors of this book chapter have initiated a number of R&D projects which main focus was to explore the concept of virtual laboratories for cement-based materials in real-life settings. A number of (journal) papers have been published about the findings of these projects in the past [4-7]. The main focus of this book chapter will be on a relatively new concept for establishing virtual laboratories which is based on a 'netcentric' approach.

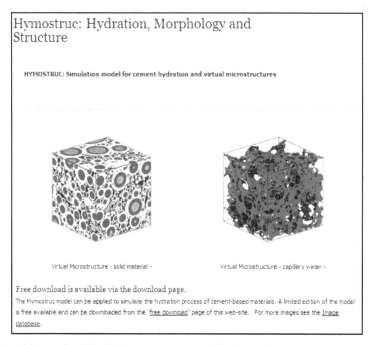

Figure 1. One of the results of the R&D projects conducted by the authors in the past.

2. The concept of netcentric virtual laboratories

Virtual experiments (or testing) are rapidly emerging as a key technology in civil engineering. Although some applications of virtual experiments other than related to materials and components have been reported by a number of researchers, most effort has been put into the development of virtual laboratories for composite materials and components. In this respect, virtual experiments are often defined as a concept of making use of high performance computers in conjunction with high quality models to predict the properties and/or behavior of composite materials and components. Consequently, virtual laboratories are often seen as a new terminology for computer simulation, which is a wrong assumption. Although it is true that computer simulation is an important tool for virtual experiments, it is only one of key components that constitute a virtual laboratory. This can be explained by the limitation of the existing composite material models. As discussed by Garboczi et al, an ideal model of a composite material or structure should be one that starts from the known chemical composition of the composite material [8]. Beginning with the correct proportioning and arrangement of atoms, the modeling effort would build up the needed molecules, then the nanostructure and microstructure, and would eventually predict properties at the macroscale level. Such fundamental and multi-scale material model, however, is still a long way off. Each existing material model has its range of applicability and its own restrictions. Corresponding computational models and supporting computer tools are developed at a number of different research institutes worldwide. However, most existing virtual laboratories are setup as web applications that only provide access to a 'closed' virtual laboratory that contains a number of (integrated) computational models and supporting computer tools.

A good example of a closed virtual laboratory is the Virtual Cement and Concrete Testing Laboratory (VCCTL) from the National Institute of Standards and Technologies (NIST) in the United States. The main goal of the VCCTL project was to develop a virtual testing system, using a suite of integrated computational models for designing and testing cement-based materials in a virtual testing environment, which can accurately predict durability and service life based on detailed knowledge of starting materials, curing conditions, and environmental factors. In 2001, an early prototype (version 1.0) of the VCCTL became public available and accessible through the Internet. The core of this prototype was formed by the NIST 3D Cement Hydration and Microstructure Development Model (CEMHYD3D). Using the web-based interface of the VCCTL, one can create an initial microstructure containing cement, mine mineral admixtures, and inert fillers following a specific particle size distribution, hydrate the microstructure under a variety of curing conditions and evaluate the properties (e.g. chemical shrinkage, heat release, and temperature rise) of the simulated microstructures for direct comparison to experimental data. As the VCCTL project proceeded, the prediction of rheological properties (viscosity and yield stress) of the fresh materials and elastic properties (elastic modulus, creep, and relaxation) of the hardened materials were incorporated into the VCCTL resulting in the release of version 1.1 (latest) of the VCCTL in 2003 [3].

In order to cope with this particularity of distributed material (computational) models, the authors adopted a relatively new concept for establishing virtual laboratories which is based on a 'netcentric' approach. In this respect, a netcentric virtual laboratory is considered as a part of an evolutionary, complex community of people (users), devices, information (i.e. experimental data) and services (i.e. computational models and supporting computer tools) that are interconnected by the Internet. Optimal benefit of the available databases containing experimental data, computational models and supporting computer tools, that replace the physical laboratory equipment, is achieved via a distributed virtual laboratory environment and is assessable for students and researchers (see Figure 2).

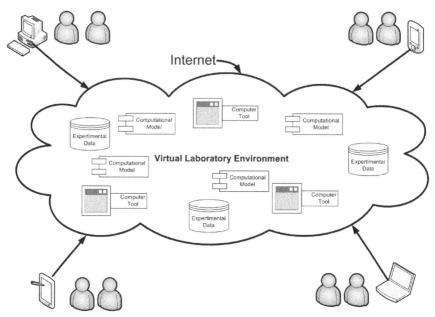

Figure 2. Virtual Laboratory Environment populated by devices, users, computational models and computer tools and databases with experimental data which are interconnected by the Internet.

3. Overview of emerging and enabling technologies

As discussed in the previous paragraph, a virtual laboratory is no longer regarded as an isolated web-based application, but as a set of integrated devices (and supported infrastructures), computational models and supporting computer tools and databases containing experimental data that, used together, form a distributed and collaborative virtual laboratory environment for virtual experiments. Multiple, geographically dispersed (research) institutes will use this virtual laboratory environment to establish their own virtual laboratory to perform experiments as well as share their result from their R&D projects. As discussed in [5], emerging and enabling technologies should fundamentally concern about the integration (or interoperability), connecting devices, computational models, computer tools and data stores. In order to structure the discussion in this section, a conceptual scheme of the different levels of 'integration' is presented in Figure 3.

3.1. Integrated infrastructure

Concerning the issue of the 'integrated infrastructure' two enabling and emerging technologies should be mentioned: Cloud Computing and Grid Computing. According to Foster et al. [9], Grid Computing and Cloud Computing are closely related paradigms that share a lot of commonality in their goals, architecture, and technology. According the National Institute of Standards and Technology (NIST) Cloud Computing (and to a large extend Grid Computing) can be defined as "a model for enabling convenient, on-demand

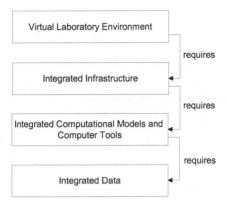

Figure 3. A Virtual Laboratory Environment requires an Integrated Infrastructure which on its turn requires respectively Integrated Computational Models and Computer Tools and Integrated Data.

network access to a shared pool of configurable computing resources (e.g., networks, servers, storage, applications, and services) that can be rapidly provisioned and released with minimal management effort or service provider interaction" [10]. In addition, Grid Computing technology adds the concept of 'virtual enterprise'. A virtual enterprise is a dynamic collection of institutes together in order to share hardware and software resources as they tackle common goals. As discussed in [3], a shared access to resources and the use of these resources is an inevitable condition for making multi-scale modeling successful.

Cloud Computing and Grid Computing also share some limitations, namely the inability to provide intelligent and autonomous services, the incompetency to address the heterogeneity of systems and data, and the lack of machine-understandable content [11]. Mika and Tummarello (2008) identified the root cause of these limitations as the lack of 'Web semantics' [12]. These limitations should be addressed at service and data levels as discussed in the next sections.

3.2. Integrated computational models & computer tools

Traditionally, the programming environment at research institutes is dominated by (programming) languages such SQL for storing, retrieving and manipulating data, Fortran, C, and C++ and Java for implementing computational models and HTML for developing web-based interfaces for end-users. Web Services fundamentally concern about the integration of computer programs, especially when the computer programs concerned are developed using different programming languages and computer operating platforms. Web Services standards and technologies offer a widely adopted mechanism for making computer programs work together.

Currently, the two main players on the web services market are Oracle (by the ownership of Sun), with their Java platform and Microsoft with the .NET platform, where both agree on the core standards (e.g. SOAP, WSDL, UDDI and XML), but disagree on how to deliver the potential benefits of Web Services to their customers. Simple Object Access Protocol (SOAP) is the standard for web services messages. Based on XML, SOAP let web services exchange information over HTTP. In addition, WSDL (Web Services Description Language) is an XML-based language for describing web services and the way how to access them. UDDI (Universal Description, Discovery, and Integration) is an XML-based registry for web services to list themselves on the Internet. XML (EXtensible Markup Language) has replaced HTML as de defacto standard for describing defining and sharing data on the Internet. The most important advantages of XML are: (1) its separation of definition (content) and representation (mark-up), and (2) its ability to support the development and use of domain specific XML vocabularies. Using the web service standards SOAP, WSDL and UDDI make computational models and supporting computer tools web services that can be accessed, described and discovered. Using XML as a basis for sharing data on the Internet will solve the interoperability problems at data level as described in the next section.

3.3. Integrated data

As discussed earlier, one of the main causes that Cloud Computing and Grid Computing share some limitations is the lack of 'web semantics'. Using XML will not solve the problem entirely; it has also its own limitations. The limitations of XML were solved by the introduction of the Semantic Web in 2004. Driving the 'Semantic Web' is the organization of content specialized vocabularies, referred to as 'ontologies'. In this respect, ontology is a collection of concepts (or terms) and constructs used to describe a particular knowledge domain. Build upon RDF (Resource Description Framework) and XML and derived from DAML/OIL, OWL (Web Ontology Language) has become the default standard for creating ontologies.

Building ontologies for describing and exchanging data between computational models and supporting computer tools in virtual laboratory environment will result in a number of different ontologies. In this respect, three different types of ontologies can be distinguished: (1) high-level (reference) ontologies that hold common concepts which can be applied for all knowledge domains and hold high-level constructs for defining the relationships between these knowledge domains, (2) knowledge domain (reference) ontologies which hold the concepts and constructs that are common within one specific knowledge domain (i.e. referring to the different macro-, meso-, micro- and nano-scale levels that exist in material research) and (3) application ontologies that hold detailed information about concepts and constructs which form the basis for sharing data between a 'group' of computational models and supporting computer tools on the Internet. From a modeling point of view, each application ontology is an extension of one or more knowledge domain ontologies, while each knowledge domain ontology is an extension of one or more high-level ontologies. Together, they form a so-called 'ontology network' that evolutionary changes over time.

4. A multi-scale modeling approach for concrete materials

The development of a virtual laboratory for construction materials requires many years of research to find out the basic principles of which a virtual laboratory should comply with, and also to find out the conditions at which a virtual laboratory would be attractive to researchers, students and people from the industry [13]. At the Delft University of Technology, first trials were focused on the virtual testing of a concrete compressive strength test, where the hydration conditions and the fracture behavior where evaluated with emphasis on the upscaling of the simulation model results. In physical-based concrete laboratories, the compressive strength is determined with an experimental device where a concrete cube is positioned in between two steel plates and compressed using a hydraulic force (Figure 4, left). The force imposed to the concrete cube is increased until failure occurs. Later, the focus was widened to other concrete properties such as tensile strength (Figure 4, right), elastic modulus, hydration temperature, etc.

Figure 4. Experimental testing device for concrete compression (Left) and tension (Right).

In a virtual laboratory, testing procedures and methods have to be mimicked using computer simulation models. These computer models are applied to simulate the different characteristics of cement-based materials and produce results that may be validated with experimental data. Especially for heterogeneous materials such as concrete materials, characteristics are modeled at different scale-levels requiring the modeling approach to be multi-scale. This means that models, operating at scale length, have to communicate with each other and exchange information by means of parameters passing which require upscaling algorithms. In general the following scale levels can be distinguished that associate with the characterization of specific materials properties (Table 1).

Scale	Property	Size range [m]
Macro	Rheology / Mechanical / Cracking / Volume stability / Durability	10^{-1} - 10^{2}
Meso	Compressive and Tensile strength / Fracture energy and Toughness	10^{-5} – 10^{-1}
Micro	Hydration / Chemistry / Pore pressure / Permeability	10^{-6} – 10^{-5}
Nano	C-S-H analysis, Calcium leaching / CH, Ca/Si -ratio / Al/Si – ratio	10^{-10} – 10^{-6}

Table 1. Overview of modeling scales, properties and size ranges.

The input for the different scale level models consists either of direct user input or of input achieved from a lower or higher scale level. This will lead to a refinement of the simulation predictability and will lead to a system with an increased synergy. A virtual laboratory can help to facilitate this linking of models and provides the opportunity to allow other partners to link their models as well, leading to an overall modeling platform for computational materials design (see Figure 5).

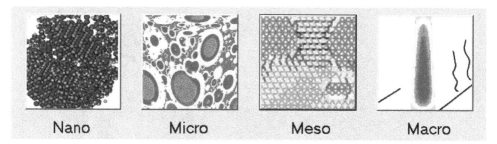

Nano   Micro   Meso   Macro

Figure 5. Schematic representation of the modeling levels.

4.1. Macro-scale level

The macro-scale level is the level at which full-scale structures are being designed and calculated. Relevant concrete related issues that have to be known for this particular scale level are listed in Table 1. For the macro-scale level, Finite element models (FEM) are very often used to simulate the structural respond of systems under static and/or dynamic actions, but also to simulate the early age hardening behavior of freshly cast material. Simulation models are used to avoid cracking during hardening. Practical problems are often related to the mix design and to the climatic conditions under which hardening takes place. Design and engineering of concrete structures, therefore, can be conducted with macro-scale FEM models like DIANA [14], ANSYS [15], ABAQUS [16] or FEMMASSE [17], where the latter is especially designed for early age analysis of concrete structures. In particular when considering the early age behavior, the development of the materials properties becomes very relevant. FEM models need to receive this information as input, or sometimes simple mathematical formulas or empirical functions are developed that can be fitted to experimental data and, with this, are able to predict the materials behavior. The disadvantage of this approach is that still expensive and time consuming physical laboratory experiments have to be conducted for the fitting and validation of the models. With these FEM models the stress and strength development can be calculated for full-scale concrete structures during the early stage of hardening. For accurate assessments, the model requires input from the structure's geometry and formwork characteristics, the ambient conditions and the thermo-mechanical properties of the hardening mix. From these inputs, the model calculates the temperature field and, from this, the development of the tensile stresses that occur as a result of the internal and external restraint (Figure 6). Important material properties herein are the development of the tensile strength, the elastic modules and the relaxation coefficient. All these properties are depending on the mix composition and on the hardening conditions, and can be expressed as a function of the degree of hydration.

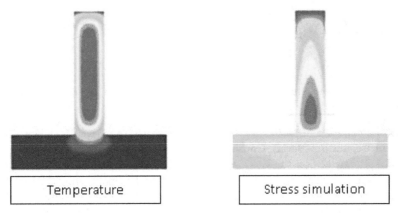

Figure 6. Temperature and stress field during early hydration of a hardening concrete wall cast on an already hardened slab [17].

With the calculated stress field and the tensile strength development known at any location in a structure, the probability of (macro) crack occurrence can be calculated using statistical methods. Cracking will occur if the calculated stress exceeds the strength while accounting for the scatter of the material properties. The accuracy of the crack predictions also requires an accurate description of the material properties. Experienced-based models, databases or numerical simulation models operating at a more detailed scale level can be used for this. More fundamental models, based on thermo-mechanical-physical mechanisms operating at an increased level of detail, can also be applied. In the next sections, models operating at the meso, micro and nano-scale level will be discussed with emphasis on the increased level of detail for the simulations while aiming to improve the prediction accuracy of the models that operate at the macro-scale level.

4.2. Meso-scale level

Increasing the level of detail of numerical schemes with the objective to simulate fracture propagation processes in concrete, lattice models can be applied. In order to be able to predict the ultimate capacity of a virtual concrete sample by simulating its cracking pattern, a fracture mechanics model is required which can handle the crack propagation of the material while loaded. A model that can be adopted in this respect is the Delft Lattice model, which has originally been developed by Schlangen and Van Mier in 1991 [18]. The model simulates fracture processes by means of mapping a framework of beams to a materials meso structure. The basic principles of the Lattice model are schematically shown in Figure 7. In this figure (a) is a schematization of a regular type of framework mesh that can be used to simulate the meso-level structure of brittle materials, such as concrete.

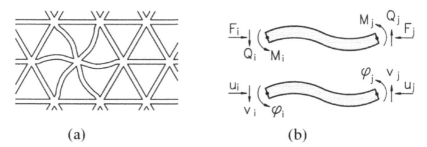

(a)   (b)

Figure 7. Principle of the Lattice model [18]; a) Lattice framework, b) Lattice beams with action forces and displacements indicated.

For composite materials in particular, the meso-level structure reflects the schematization of the paste phase, the Interfacial Transition Zone (ITZ) and the aggregate explicitly. This approach of schematization fits very well with the level that is required for modeling the compressive stress calculations inside a virtual laboratory. The model should be able to detect failure paths through the material (weakest links) and to calculate the accompanying ultimate strength of the building material from it. Once this failure path has been initiated, the inner structure of the material starts to disintegrate and the strength capacity will reach its maximum. After having reached this maximum strength level, a descending branch will follow that indicates the post-peak behavior of the material. The Lattice model is capable to calculate this part of the failure traject and to quantify the fracture energy of the failure behavior as well. For conventional concretes, the Interfacial Transition Zone (ITZ, weak bonding zone around the aggregates) is almost always the weakest part of the material that initiates and contributes to the failure paths (Figure 8). For higher quality concretes, the failure paths might cross through the aggregate particles which implicitly affect the brittleness of the material. A proper compressive strength model within a Virtual Laboratory should therefore implicitly deal with these different kinds of failure mechanisms related to the mix composition in general and the inner microstructure of the material in particular.

4.3. Micro-scale level

For the simulation of the evolving cementitious microstructure that forms the fundamental basis for the development of the material properties the hydration model Hymostruc can be used [19,20]. After mixing, hardening commences and the material properties start to develop. This process leads to a set of properties that is unique for every particular type of

 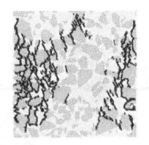 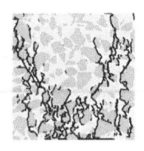

Figure 8. Left: Concrete crack pattern after loading. Right: Lattice simulation [18].

cement-based material. The Hymostruc model (Figure 9, left) can be applied to predict the actual state of the material properties with the degree of hydration as the basic parameter. The model calculates the hardening process of cement-based materials as a function of the water-cement ratio, the reaction temperature, the chemical cement composition and the particle size distribution of the cement. The model calculates the inter-particle contacts by means of the 'interaction mechanism for the expanding particles' (Figure 9, right) where hydrating particles are embedded in the outer shell of larger hydrating particles. This mechanism provides the basis of the formation of a virtual microstructure which, on its turn, can be considered as the backbone of the evolving strength capacity of the material. When considering the virtual laboratory, the Hymostruc model will operate at the micro-scale level and will be used to calculate the internal microstructure that is necessary to simulate the compressive and tensile strength development, the development of the elastic modulus and other microstructure related properties. The microstructure of the material can be considered as the morphology-based inner structure of the paste, i.e. the "glue", that tightens together the aggregate particles and/or other composite phases such as fibers, fillers, etc, inside a composite material. Failure of the paste structure, therefore, strongly depends on the strength characteristics of the internal bondings in the microstructure of the paste. Modeling the morphologies of these bondings in terms of their chemo-physical nature has to be resolved at the nano-scale level. The development of the properties of the C-S-H gel, therefore, is a scale-level that has to be considered as well.

4.4. Nano-scale level

Nano-scale modeling has benefit from an enormous increase in the attention of the research community with the aim to model the chemical and physical based processes of the Calcium-Silicate-Hydrates (C-S-H gel) that forms the fundamental elements of the hydration products of cementitious materials [21]. Characterizing the materials performance at this particular scale level asks for modeling the fundamental processes using molecular dynamics principles. For cement-based materials in particular, emphasis has to be on the characterization of the basic building blocks of the C-S-H nano-structure that operate at the sub-micro scale level. This intermediate scale-level between the nano and micro-scale level enables a modeling approach that bridges the gap between the micro and nano level and that enables an exchange of fundamental materials properties (Figure 10).

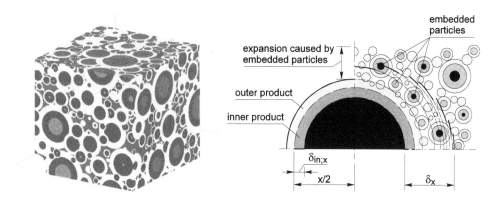

Figure 9. Left: 3D virtual microstructure simulated with Hymostruc. Right: Hymostruc interaction mechanism for expanding particles representing the formation of structure of the virtual microstructure [19,20].

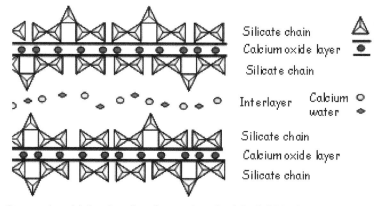

Figure 10. Structural model that describes the atomic scale of the C-S-H gel [21].

4.5. Up-scaling

The development of numerical algorithms that allows for particular scale-level information to be used at other scale levels is the most challenging part of multi-scale modeling. Bridging the length scales between the nano scale level and the macro-scale level requires a upscale models that enable to span of 10 orders of magnitude (see Table 1). In this approach the nano scale level forms the basis of the multi-scale framework. The output properties calculated at a particular scale level forms the input at a higher scale level. This approach enables the analysis and design of composite materials starting from the fundamental nano-scale level and evaluates the results at the full-scale macro-level. It opens the door for tailor made design of composite materials and is a first step towards property defined modeling approach. A virtual laboratory is an excellent vehicle to achieve this.

5. Virtual laboratory prototype

In 2012, the first prototype of a virtual laboratory has been developed at Delft University of Technology. In this paragraph, the system development rationale and architecture are presented, including screenshots of the prototype created to demonstrate the idea of a web-based virtual laboratory. Due to the fact that existing computation models and supporting computer tools have been implemented in the traditional way in the past, the demonstrated prototype not fully operates according the netcentric approach as discussed above. The main purpose of this prototype is to support the approach of multi-scale modeling for concrete materials in an integrated web-based environment. From this, it can be derived that prototype should provide a complete environment for multi-scale experimentation easing the study of composite materials.

From the point of view of the final users, the prototype should aid and support them on their experiments relying on a fast execution of the best simulation models existing. Since the system's user interface is based on the functionalities available in the computational (simulation) models and supporting computer tools, the final user needs to know about the computational models to be able to work with them. However, this user does not want to care where these computational models are running – they want to focus on their educational and research applications instead of the technology required providing their needs. For instance, the system should relief the users from the burden of looking for simulation module availability, its installation and execution; take care of the issues regarding combining modules to perform multi-scale experiments and carry out other system administration duties too. From the point of view of the computational model creators, the virtual laboratory should be open in a way to be able to support new computational models to be plugged-in adding new services and functionalities to the existent ones. The computational models must exchange data, making possible the execution of a multi-scale experiment based on different computational models developed independently (i.e. mashup). Therefore, there must exist a platform that works as an open ecosystem environment, populated with simulation models, created and executed independently, but that are coordinated by the final users through an interface that offers multi-scale experiments of composite materials. With this purpose, the system architecture was created based on two principal modules, the user interface module (frontend) and the simulation modules (backend). This modular organization provides decoupling of the user interface from the simulation modules. The deployment diagram of the virtual laboratory system is presented in Figure 11. showing the fronted module as the Graphical User Interface (GUI) component and the backend module as the model execution controller, which relies on a Grid infrastructure to execute the computational models.

The virtual modelling laboratory contains a rich interface web application that has been developed using cutting-edge technology, such as HTML5, CSS3 and JavaScript. The backend is defined as a set of web services which are developed as a Restful API and which is accessed through AJAX calls. An extensive set of toolkits and frameworks is used in the final implementation. The GUI foundation is based on the Twitter Bootstrap library for UI building and on the Backbone.js toolkit for organizing the JavaScript code, and being fully based on the jQuery library. The backend was developed using the Play framework. Although the basic application architecture is defined as a two-layered system – frontend and backend, the backend layer is composed by a set of different computational modules that can be seen as independent web services. The modules executions are managed coordinated by the overall controller service, but the composition architecture is much more complex than what is exposed. Different researchers and developers, distributed over the Internet, can make their computational models available through different interfaces. In addition, the computational models can also use results from other computational models.

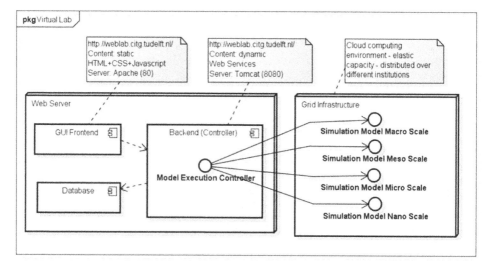

Figure 11. The deployment diagram of the virtual laboratory at Delft University of Technology.

As discussed earlier, the prototype relies on a Grid infrastructure to execute the computational models. This infrastructure is distributed over different institutions, because each computational model owner should be responsible for the model execution and availability as a web service. For this purpose, a Grid infrastructure should be available to support the whole platform. As discussed earlier, other new emerging technologies such as Cloud Computing can supply on-demand computational resources to run the computational models.

In order to be able to transfer data between the different modelling scale levels, multi-scale modelling principles have to be considered as well. Apart from how the data is transferred from one scale level to another, the most challenging part is how to connect the levels from a modelling point of view. Bridging the scale levels can go along with transfer of data only by means of parameters passing or by means of a more complicated integration of models that operate at different scale levels or a possible combination. In either way, bridging the scale levels is an intensive modelling work that requires significant effort both from materials properties as well as from a modelling point of view. With the virtual modelling lab, the user can choose at which level he/she wants to start the numerical experiments (Figure 12). It is organized in such a way that data can be calculated in a certain scale level and, when the user decides to proceed with an analysis at another scale level, the data can be taken.

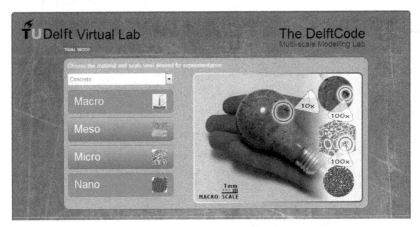

Figure 12. DelftCode: Selection of scale level.

The pilot version of the virtual laboratory at the Delft University of Technology (referred to as DelftCode) provides output (results) which are represented as graphs in the GUI. In addition, each simulation of computational model at a certain modelling scale produces specific parameter output that is managed by the DelftCode framework in such a way that it can be used as input for computational models at other scale levels (Figure 13). In this way active multi-scale modelling can be conducted and the results can be reused at other scale levels. The way how it is implemented in the DelftCode framework is that after conducting numerical experiments the user can switch to other scale levels. For the other scale levels the same procedure is followed. Since data will be stored and available for all scale levels, the user can access data of previous numerical experiments and reuse it for other experiments.

 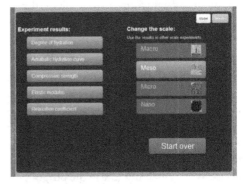

Figure 13. DelftCode: Model setup and result outputs.

At time of writing, the prototype only supports computational models at micro and meso scales, but the prototype has been designed and implemented to support all scale levels. Figure 14 show all the proposed user interactions in the prototype. These interactions show the possibilities of the developed prototype that allows multi-scale modeling platform. The architecture of the prototype is designed in such a way that future extensions in terms of new model additions or adding another scale level can easily be achieved.

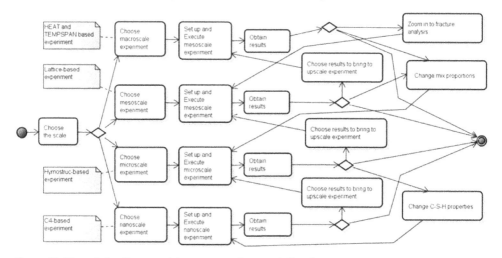

Figure 14. The activity diagram of the user experiments of all scales.

6. Conclusion and discussion

This paper depicts the development the concept of so-called virtual laboratories that are based on a netcentric approach. In this context, a netcentric virtual laboratory is considered as a part of an evolutionary, complex community of people (users), devices, information (i.e. experimental data) and services (computational models and supporting computer tools) that are interconnected by the Internet. A large number of emerging and enabling technologies are discussed which form technological basis for establishing such netcentric virtual labs. Although the developed prototype is hampered by the fact that most computational models and supporting computer tools have been developed using traditional programming languages and platforms, it showed the enormous potential of the netcentric approach advocated in this paper.

From the perspective of the application domain, a virtual laboratory can be considered as a most appropriate way to interact with users and developers of computational material models at different scale levels. In this paper the approach of multi-scale modeling approach for concrete materials is explained. This approach is based on numerical (computational) models developed for cementitious materials that operate at different 'geometrical' scale levels. With the ability to use the output generated at any particular scale level as input for models that run at other scale levels, the web-based virtual laboratory acts as a real multi-scale modeling platform. The architecture of the proposed virtual laboratory as provided in this chapter allows the models to be exchangeable and merge-able leading to an integrated approach.

The numerical models for simulating the materials performance operate at the different scale-levels with the Hymostruc model as the main microstructural model at the micro-scale level. With this model connecting to the nano-scale model for inputting detailed information on C-S-H gel properties, the microstructural information can be used as input for the meso-level Lattice model to simulate the fracture behavior of composite materials submitted to internal actions (drying, autogenous shrinkage, etc) or external actions (loads, thermal imposed loading, etc). These models can generate input data for the macro-scale models to simulate the full scale performance of structural elements. With the multi-scale approach, the consequences of changing parameters that act as input for lower scale models (nano, micro, meso) can directly be made visible by upscaling. Therefore, from this approach the following conclusions can be drawn:

- The netcentric virtual laboratory is a most appropriate tool for the assessment of composite materials performance using a multi-scale modeling approach;
- The web-based approach enables the communication between models that operate at different geometrical scale-levels using an integrated computational modeling system;
- The prototype shows the huge potential of web-based modeling and provides an exchangeable and scalable system for multi-scale modeling;
- The future perspective of virtual web-based modeling shows to be a very powerful alternative for vast computational models that run over different time and length scales.

Author details

E. Dado
Netherlands Defence Academy, Breda, The Netherlands

E.A.B. Koenders
Delft University of Technology, Delft, The Netherlands,
COPPE-UFRJ, Programa de Engenharia Civil, Rio de Janeiro, Brazil

D.B.F. Carvalho
(PUC-Rio), Rio de Janeiro, Brazil

7. References

[1] Kuester, F. and Hutchinson, T. A virtualized laboratory for earthquake engineering education, ASEE Journal of Engineering Education, 15:1, 2007.

[2] ECWINS consortium. The Road to Standardized Window Production. Collective Research Projects for SMEs, Vol. 3, European Union, 2007.

[3] Bullard, J. et al. Virtual Cement, Innovations in Portland Cement Manufacturing, USA, 2004.

[4] Dado, E., Koenders, E. and Mevissen, S. Towards an advanced virtual testing environment for concrete materials, published in the proceedings of the MS2010 conference, 2010.

[5] Dado, E, Koenders, E. and Beheshti, R. Theory and Applications of Virtual Testing Environments in Civil Engineering, International Journal of Design Sciences and Technology, 16:2, 2009.

[6] Koenders, E., Schlangen, E. and Dado, E. Virtual testing of compressive strength of concrete, published in the proceedings of the ISEC-4, Conference, 2007.

[7] Koenders, E., Dado, E and van Breugel, K. A Virtual Environment for Multi-Aspect Modeling, published in the proceedings of the SCI2004 conference, 2004.

[8] Garboczi, E., Bullard, J. and Bentz, D. Virtual Testing of Cement and Concrete, Concrete International, No. 12, United States, 2004.

[9] Foster, I., Zhao, Y., Raicu, I. and Lu, S. Cloud Computing and Grid Computing 360-Degree Compared. Proc. Grid Computing Environments Workshop, p.1-10, 2008.

[10] Mell, P. and Grance, T. The NIST Definition of Cloud Computing, NIST special publication 800-145, United States, 2011.

[11] Wu, Z. and Chen, H. From Semantic Grid to Knowledge Service Cloud, Journal of Zhejiang University, 13:4, China, 2012.

[12] Mika, P. and Tummarello, G. Web semantics in the clouds. IEEE Intelligent Systems Magazine, 23-5, United States, 2008.

[13] Koenders, E, Schlangen, E. and Dado, E. Virtual Testing of Compressive Strength of Concrete, Proceedings of the ISEC-4 Conference, Australia, 2007.

[14] DIANA, http://tnodiana.com/.

[15] ANSYS, www.ansys.com/.

[16] ABAQUS, www.simulia.com.

[17] FEMMASSE, www.femmasse.nl.

[18] Schlangen, E. Experimental and Numerical Analysis of Fracture Processes in Concrete, PhD thesis, Delft University of Technology, The Netherlands, 1993.

[19] Breugel, van, K. Simulation of Hydration and Formation of Structure in Hardening Cement-based Materials, PhD. Thesis, Delft University of Technology, Delft, The Netherlands, 1991.

[20] Koenders E. Simulation of Volume Changes in Hardening Cement-Based Materials, PhD-Thesis, Delft University of Technology, Delft, The Netherlands, 1997.

[21] Dolado J. S., Hamaekers J. and Griebel M. A Molecular Dynamic study of cementitious Calcium Silicate Hydrate (C-S-H) gels, Journal of American Ceramic Society.

Chapter 11

Advanced Composite Materials by Resin Transfer Molding for Aerospace Applications

Susanna Laurenzi and Mario Marchetti

Additional information is available at the end of the chapter

1. Introduction

Competitiveness drives the aerospace industries to investigate new technology solutions to address market pressure and high-tech demands. The global objective is to reduce to half the amount of fuel by 2020 and at least 70% less by 2025 with respect to the Boeing 777, one of the most efficient aircraft, which is made entirely of carbon fiber. The weight saving to increase payload and the reductions of the cost/time of the production cycle are imperative targets. For these reasons, aerospace companies, which are traditionally based on the use of metal alloys, have been focusing for past decade on composite materials. The main advantages of composites with respect to metals, that are resistance to corrosion and fatigue and high performance/weight ratios, are a set of qualities for winning the current and future aerospace applications. Obviously, this is possible only through the development of economically competitive technologies.

The Resin Transfer Molding (RTM) is one of the most promising technology available today. RTM is capable of making large complex three-dimensional part with high mechanical performance, tight dimensional tolerance and high surface finish. A good design by RTM leads to fabricate three-dimensional near-net-shape complex parts, offering production of cost-effective structural parts in medium-volume quantities using low cost tooling. In addition to these advantages, the problems of the joints, typical of the metal structures, can be eliminated by integration of inserts.

The final performances of a composite depend not only on the choice of the matrix and the fiber but also on the manufacturing process by which they are made. Since the starting of the composite life, the presence of imperfections due to manufacturing must be considered. Such imperfections can be already damage for the manufactured composite piece or lead to the damage quickly. The damage for composites can be defined as a change in the microstructure of the material that causes deterioration in the structural behavior of the component and sometimes its collapse. The damage in a composite structure can occur at the level of fibers and matrix as well.

The most common damage to the fibers is the interruption of their continuity. For example, fibers that are subject to load tend to align again inducing states of compression and tension on the matrix. These states may cause, in addition to a local decrease of the properties of the lamina, the breaking of the fibers themselves and the gap between fiber and matrix. In the assumption of an optimal stratification, fiber misalignment is caused during the manufacturing process by a bad balance of the process parameters that leads to the deformation of the fiber bundles. The fibers may also be distributed unevenly in the volume of the composite, and this generates intra-laminar shear stresses under operating conditions.

In the matrix, the damage is essentially correlated to the presence of porosity. The formation of microvoids between fibers and dry spots are potential starting points for propagating cracks or delaminations. The response of the component to delamination depends not only on the compatibility and surface tension between fibers and matrix but mainly on the compaction, and therefore also on the impregnation phases during the manufacturing process.

Comparing RTM with traditional manufacturing process applied in the aerospace industries (i.e. autoclave), the RTM technique results in a suitable alternative to the prepreg approach permitting high finish quality and controlled fiber directions. RTM reduces voids compared to hand lay-up so increasing component mechanical properties. In addition, component design by RTM process can compete with metal one when prepreg cannot be applied to manufacture a product. Hand lay-up requires a low initial investment, but it becomes more expensive (cost per product) than the other techniques due to recurring costs associated with direct labor and material waste. Compared with compression molding (CM), RTM requests lower tooling costs because of the absence of a press system to compact the preform. RTM seems to meet both low cost/high volume requirements of the automotive industry (500 to 50,000 parts per year) and low number/high performances (50 to 500 part per year) of the aerospace industry. In fact, RTM can guarantee the demanded performances for the aeronautical production: reduction of the mass, increase of the operating life, aimed design, reduction of the production times. Today an increasingly number of parts realized using RTM is observed due to the development of new resins and the preforming technology. In addition, such technique is suitable, with only small adjustments, for the realization of large, complex and thick-walled structures for use in infrastructures and military applications.

2. Overview of RTM process

The RTM is a process with a rigid closed mold. Figure 1 summarizes the main steps for a simple case. The lamination sequence (preform) is draped in a half mold, then the mold is closed and the preform compacted. After that, the resin is injected using a positive gradient pressure through the gate points replacing the air entrapped within the preform. Usually, vacuum is applied at dedicated vents in order to favorite the air escape from the mold. When the resin reaches the vents, the gates are clamped and the preform is impregnated. At this point, the cure phase is considered to start. Finally, the mold is opened and the part removed. Especially for aerospace structures, an additional free-mold post-curing phase can be necessary in order to guarantee the polymerization of the matrix and release the internal thermal stress.

The closing mold step is characterized by the compaction of the fiber reinforcement, which permits to reach the desired thickness and design fiber volume fraction. The compaction changes the microstructure and the dimensions of the preform, producing large deformations and nonlinear viscoelastic effects. These effects are accompanied by a change in energy within the material, which causes the residual stresses due to the viscoelastic behavior of the fibers. However, during the impregnation phase a release of stress, probably due to the balance, occurs.

The injection phase must guarantee the complete impregnation of the preform: a bad impregnation of the fibers results in dry spot areas with missing adhesion between the layers, which makes the surface rough and irregular. If partial impregnation occurs in the proximity of a connecting zone among elements, it can cause a bad integration with a consequent loss of mechanical properties.

MAIN STEPS OF RTM PROCESS

Figure 1. Sequence of the main steps of RTM process. From left to right: realization of the preform, deposition and draping of the preform in the mold cavity, closing mold, injection phase and curing phase; after the curing phase, the mold is opened and manufacred element is demolded.

2.1. Process parameters

The RTM process is governed by variables and parameters that are dependent on each other. Their combination affects the process and the quality of the finished product. Consequently, they need to be carefully determined. The most important parameters, which can not be neglected in the design, are pressure, temperature, viscosity, permeability, volume fraction, and filling time of the process. There are also a multitude of parameters that must be considered independently, such as the angle of attack of the nozzle, the orientation of the fibers, the paths of flow and shear rates, the stratification. In fact, the resin tends to flow more quickly in the fiber direction, thus the flow dynamic depends mainly on the type of fabric used and the number of overlapped layers. Sometimes it may be necessary to have a certain number of skins, not for structural reasons, but to obtain a homogeneous distribution of the resin. The thickness of the part to be manufactured can also affect the flow progress and the impregnation of the fibers, causing a high percentage of voids and dry spots. The thickness becomes a critical design constraint especially in the case of the inclusion of reinforcements and ribs.

The injection pressure determines the injection velocity of the resin into the mold, the hydraulic pressure and the holding and closing forces of the mold. Consequently, the injection velocity defines the filling time, which should not be too short to ensure an adequate impregnation of the fibers and, at the same time, the filling must be such as to avoid the risk of incurring in premature gelation of the resin. The injection pressure adjusts the distribution of the resin on the preform, which affects the formation of air voids in the matrix, the appearance surface and the mechanical properties of the finished product. Another phenomenon in which this parameter is relevant, together with the viscosity, is the so-called "fiber wash", i.e. the movement of the reinforcement inside the mold during the injection phase. In this case, the surface treatment of the fibers and especially the choice of the binder play a fundamental role. If the binder dissolves too quickly in contact with the resin, then fibers under the injection pressure can move freely.

The temperature is an extremely important process parameter and it is strictly related to the injection pressure and the viscosity of the resin. When the temperature increases, the filling time decreases and the working pressures are lower. When the temperature is low, the viscosity of the resin increases and it is necessary to increase the pressure to ensure the transfer of the resin itself.

3. Preform technology

As previously mentioned, the diffusion of RTM process in the aerospace industries is strictly related to the possibility of building net-shape 3D-complex structures. This possibility is given by the development of the preform technology. Preform is prepared separately and constitutes the skeleton of the final product, greatly simplifying the molding operations and reducing the time and cost of processing. When the production volumes is medium-high, an industrialized preform permits to amortize in a relatively short time the cost of equipment, even in a limited production, as required in the aerospace industry.

The textile methods are numerous. The choice of one method over the other depends on several factors: the processability, the feasibility of the geometry, the desired mechanical properties of the molded part, and the cost of production. Obviously, the preform affects strongly the performance required by the final application. The choice of the architecture of the fiber reinforcement depends on the required performance of the composite structure and the characteristics related to the process, such as permeability, compressibility and drape.

The preforms are formed weaving yarns or rovings. The terms used for the fiber comes from the textile tradition: a single fiber is a filament; a set of fibers produced simultaneously is called strand. Several parallel strands can be rolled up like a ribbon, called "roving" or "tows", or twisted and united as a strand, which is named "yarn". Generally, the yarns are typically not too complex because the excessive torque reduces the possibility of penetration of the resin in the cavities. Besides, a slight twist of the yarn compacts and raises the formation of composites with high fiber content.

Strands and yarns can be processed as woven roving and cloth. The first type uses effectively the resistance of the fibers, but it can produce composites with high resin content as the roving is not compacted. Woven roving is used to rapidly produce thick composites, because resin can fill them easily. Cloth is slightly less resistant than the roving because of the slight damage that is produced by rolling the fibers and also due to the twisting of the fibers themselves. The cloth is well impregnated from the resin and, as it is compact, high fiber content is obtained.

The weaving process can produce a wide variety of forms. The ratio between the number of filaments in the transverse direction (weft) and the number of filaments in the longitudinal direction (warp) may vary from nearly pure unidirectional, in which the number of filaments of the weft is the minimum required to hold together the fabric, up to the case of equal number of filaments in the warp and weft. In addition, there are hybrid fabrics with carbon fibers in the warp and glass fibers as weft. The weaving method influences the properties of the composite. In flat woven, warp and weft overlap in an alternating manner. In particular, weaving the warp passes through a number of fibers of the weft. These fabrics are better suitable for complex shapes, and their strength and stiffness are slightly higher because the fibers are straight on average. There are also more complex forms of weaving, with an angle of intersection fibers different from 90°. These methods are used to produce biaxial and multi-axial weaves. The complex fiber architectures can be obtained with the weaving method by interlacing and knitting the fibers along the three spatial directions; an example is shown in Figure 2. Different bi-axial layers can be also stitched. The stitching method consists on darning the layers with fibers (Figure 2). These are automated techniques that realize complex shapes and 3-D junctions in place of bolts and rivets. From the mechanical point of view, these methods can increase the resistance to fracture of the structure and reduce the progress of cracks by adopting high strength fibers in the z-direction. On the other hand, the stitching seams can produce local defects induced in the preform as a result of penetration of the wire and the needle. Further, the robotic system is very expensive and sometimes damage due to misalignment of fibers can occur.

The choice of the fiber reinforcement is carried out according to the mechanical requirements of the manufactured component. The types of fibers that can be used are many and with variable characteristics. During machining, the preform can be damaged, particularly when using glass fibers that are fragile and have a high friction coefficient (about 1 in the contact glass on glass). For this reason, immediately after production, the fibers are protected with a coating, "finish" or "size", which performs several functions: it acts as a binder to hold the fibers together, as a lubricant to reduce the coefficient of friction between fibers, and it improves the wettability of the fiber and the adhesion of the resin. Purpose of sizing is also to protect the fibers from the environment, mainly by moisture. These protective sizes are specific to each type of resin, since they must be dissolved from the resin itself. In the case of carbon fibers, the epoxy resin is also used as a protective finish.

Braiding method is another interesting textile process for aerospace structures. The braiding has high level of conformability to complex shapes and it is especially suitable for conical and cylindrical geometry. It regards 2D and 3D preforms. The high level of process automation reduces scrap and labour costs. More advantages of the braiding technique compared to standard tape and fabrics are strength in third dimension, improving fatigue resistance, more efficient distribution of mechanical stresses and the possibility to consider inserts during the preform realization. Despite of these characteristics, the braiding process requires the use of a mandrel and high initial cost for the preform engineering.

An important target of any textile process is obtaining the desired fiber volume fraction that is given by the design. It is very difficult to determine precisely the volume of reinforcement that must match the structural design with the manufacturing aspects. If the fibers are too compacted or their content is excessive, there is no sufficient space for the passage of the resin and the filling time becomes longer. Generally, the optimum volume of fibers is 60% of the product. An increase with respect to this value may cause a bad distribution of the resin and a dramatic drop of the mechanical properties of the manufactured component.

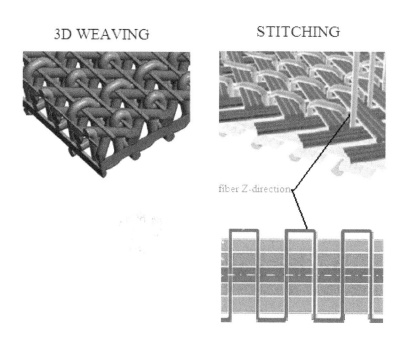

Figure 2. Examples of 3D weaving and stitching textile processes.

Figure 3. Braiding technology

4. Resins in aerospace composites

Thermosetting resins are the matrix used in RTM because of their low viscosity during the process. They are very rigid materials consisting of cross-linked polymers in which the motion of polymer chains is strongly limited by the high number of existing crosslinks. During the polymerization phase, thermosetting undergoes irreversible chemical change.

The selection of a reactive thermosetting resin for RTM application forces to deal with a large number of choices of chemical engineering, mainly due to a strong relationship between the chemistry and process engineering. In fact, the process parameters, such as temperature and pressure, cannot be selected without considering the chemistry of the resin to be used. Factors to take in consideration for a RTM system can be divided into two broad categories: processing and performance. Initial viscosity and molding life are function of the temperature, and they determine the operational temperature range of a process. The molding time is a function of the rate at which the reaction occurs between the resin and the curing agent and the rate is directly proportional to the temperature. The viscosity depends on the chemical-physical characteristics of the matrix. Viscosity may change over time because of both temperature variations and as consequence of chemical reactions that occur in the liquid state. The knowledge of the rheological behavior of systems is essential for a proper setting of process parameters. In fact, the values of the viscosity during the phases of the process must be such as to guarantee both the simultaneous removal of dissolved gases and moisture entrapped in the matrix, and the compaction of the fibers, before reaching the gel point. However, the viscosity of the resin must not be too high, especially in the case in which the fiber volume fraction is higher than 40-50%.

Within thermosetting resins, several classes are suitable for aerospace applications of RTM, such as epoxy, phenolic, cyanate and bismaleimide. Phenolic resins are produced from the reaction of formaldehyde with phenol to give the condensation products. The reaction is always conducted in the presence of catalysts: these can be both acid and bases and their choice has a determinant role, together with the molar ratio of the reactants, and the nature of the reaction products. There are mainly two types of phenolic resins: resole and the novalacs.

The base resins and formulated systems are available from Borden and Georgia Pacific. Approximately 91,000 tons of molded materials and phenolic are produced every year, but only a relatively small amount is used in composites for structural applications. A large number of industrial applications of phenolic resins are based on the excellent adhesive properties and strength. Especially when combined with appropriate reinforcements, phenolic resins have good chemical and thermal resistance, good dielectric strength and good dimensional stability over a wide range of temperatures. Materials produced with these resins have a very low flammability, are very resistant to creep, have low moisture absorption and have a remarkable resistance to degradation from a variety of lubricating fluids. Their low viscosity and the high char yield of these materials are used in many structures forming composite carbon/carbon. Phenolic resins are polymers that generate volatiles (mostly aqueous vapor) during the treatment phase. The volatile substances have a strong impact on the processability of these composite materials resulting in structural voids where the content can have a dramatic effect on the mechanical properties of the part. In addition, these polymers are generally brittle, characterized by a low elongation and a low tensile strength. Despite the relatively poor mechanical behavior and the difficulty of processability, these resins are becoming more applied in the structural field due to the low flammability and low smoke production. The only resins comparable to the phenolic for the properties of high fire resistance and low smoke generation are the bismaleimide resins, but they are at least an order of magnitude more expensive. The nature of the RTM process (closed molding) makes it difficult the implementation of this method due to the development of volatile substances typical of phenol in the curing phase. In order to overcome this problem clever methods have been developed. A particularly interesting approach is the use of a catalyst that allows gelation of the resin at temperatures below 100 °C. So if the water is formed during the RTM process, but the temperature is maintained below the boiling point (100 °C), the water will remain in the liquid phase and will act as a plasticizer. If the temperature increases above 100 °C in the initial stage of gelation in the curing phase, the water will evaporate and will leave the part producing a massive degradation. A high temperature cure achieved with a ramp rather slow through the 100 °C will lead the water out from the gelled structure and the cross-linking of the polymer occurs without degradation of the piece. Another issue to consider when phenolic resins are used in RTM is the acidic nature of many catalysts, which can cause corrosion in some materials for molds.

Epoxy resins are well known in the production of composite aerospace materials. The vast variety of epoxy and cure agents makes these systems very versatile in terms of manufacturing process and obtainable physical properties. Although in the last twenty years, a large innovative work in developing new formulations of epoxy resins was done, only a limited marketing of new resins was realized. When liquid epoxy resins such as DGBA and DGBF are used in RTM, they are usually part of a two-component systems. In this case the selection of the curing agent is really important. In the polyester and vinyl ester resins, the catalysts alter the cure time but do not have substantial effect on the viscosity and on the final properties of the polymer. In an epoxy system, the selection of the curing agent is crucial because it determines the thermal and mechanical properties of the matrix and defines the dependence of viscosity from the temperature, thus controlling the processability of the system. Epoxy resins polymerize with many materials such as polyamines, polyamides, phenol-formaldehyde, urea-formaldehyde, and acid anhydrides. The reactions taking place can be coupling or condensation reactions. DETDA is a liquid aromatic amine which is widely accepted as the primary hardener in many RTM formulations. DETDA is liquid at room temperature and provides good processability in RTM, both in a single-component system and bi-component one. The slow curing of epoxy system with DETDA allows these systems to be processed in a wide temperature range. The inclusion of a catalyst in the formulation of a DETDA significantly increases or decreases the reaction rate and lowers the cure temperature. The DETDA produces a polymer with high glass transition temperature (generally Tg > 177 °C) when fully cured. The Young's modulus is usually less than 3.1 Gpa. Therefore, the use of a liquid aromatic amine such as DETDA provides an excellent processability with high glass transition temperature but with a very low value of the elastic modulus for space applications.

The aliphatic and cycloaliphatic amines are useful agents for treatment of many epoxy resins. These materials are almost all liquids with low viscosity (as DACH, IPDA, PACM, etc.) that are readily soluble in the formulation of epoxy resins. In these cases, the thermal and mechanical properties are inferior to those obtained with the aromatic amine. With these agents, the glass transition temperature is in the range 121 °C – 177 °C, and the elastic modulus of the polymer is in the range 2.4 GPa – 3.1 GPa. Anhydrides were widely used for many years as agents for epoxy polymerization in applications of filament winding. This class of catalysts has not received much attention in RTM applications. The liquid nature of most of the commonly used anhydrides (MTHPA, NMA, etc.) and their good solubility with the epoxy indicate that they could be used in RTM monocomponent and bicomponent systems. The availability of a large range of hardeners for epoxy-anhydride makes them available for systems that can meet specific process requirements. Generally, the formed polymers have a glass transition temperature of approximately 140 °C – 150 °C and a modulus of 3.45 GPa, tensile elongation of 3% – 5% and moisture absorption of 1.5%. The curing phase with an anhydride is relatively complex. The mechanism of cure can have several important consequences, such as low moisture absorption and in many cases a high final operative temperature. When anhydrides are exposed to moisture, usually over a long period of storage in a humid environment, they form acids. These acidic components interact with the basic catalysts inhibiting the polymerization, lowering the values of the modulus and the glass transition temperature, and increasing the absorption of moisture. Epoxy systems that have the greatest success in the aerospace market are the mono-component ones. There are substantial reasons for the success of these systems, such as ease of use, quality control of both process and materials and the excellent thermal and mechanical properties. In general, epoxy resins combine incomparable properties of flexibility, adhesion and chemical resistance.

The cyanate resins have a relatively small niche in the production of composites. The very high glass transition temperature and the excellent mechanical properties of the polymer are the primary guides to their use. The relatively high cost of these systems (150 – 500 U.S. $ per kilogram) prevents the entry into any market that is not closely related to their required application. This resin is generally cured with the addition of heat and transition metals as catalysts [Co(III), Cu(II), etc.]. The gelation occurs at about 50% – 60% conversion, similarly to the chemistry of an epoxy resin. The coefficient of thermal expansion is relatively small, about 50 ppm / °C. Cyanate resins are often formulated with epoxy resins or maleimides in order to modify the processability and properties of the resulting polymer. Generally, the formulations include salts of transition metals and phenolic species as catalysts. There are very few formulations originally formulated for RTM. Usually cyanate resins produce a polymer with high transition temperature, low moisture absorption, good mechanical properties and excellent electrical properties. The use of them in the structures and radomes is driven by the needs of a high transition temperature coupled with a low dielectric constant and low dissipation factor to prevent degradation by high energy radiation transmitted and received through the structure. Some formulations have a flat response for both dielectric characteristics and dissipation in a wide range of temperatures and wavelengths of electromagnetic radiation. For these reasons, cyanate resins are widely used in aerospace fields. In particular, composites made by carbon fibers and cyanate resins are used in satellites for very rigid structures, with a high transition temperature and low absorption of moisture that can withstand to repeated thermal cycles without failure due to internal stresses.

Bismaleimide resins (BMI) currently provide a market niche in the manufacture of composite structures, i.e., for those parts which require a very high glass transition temperature, good stability and thermal oxidation and low flammability. The relative high cost of BMI resins (44 – 260 U.S. $ per kg) limits the applications to advanced ones. The BMI are produced by a reaction of aromatic bi-ammine precursor with maleic anhydride. The resulting resins are cured with heat and in general without additional catalysts. The unmodified BMI are fragile materials with failure strain less than 2%. In order to make these systems more resistant, modified mixtures with amines, monomers, vinyl or epoxy resins are formulated. The polymers formed in this way are reasonably resistant, but with the increasing of post-cure temperatures the fragility increases. A typical BMI resin is produced by the reaction of methylene dianiline (MDA) with maleic anhydride, using heat to remove the water produced and push the reaction to completion. A variety of process features can be obtained by changing the properties of the molecular backbone of BMI resin and adding co-reactants. The use of BMI is driven primarily by their exceptional performance at high temperatures, in particular by their ability to maintain mechanical properties at more than 149 °C to below the saturation moisture. The other characteristics that lead to the use of BMI are their good electrical capacity, long term stability to the thermal oxidation for higher temperatures up to 177 °C and the exceptional capacity to not generate smoke when exposed to high heat fluxes. These characteristics at high temperatures push to use these resins in the aerospace field, although at present their main application is the manufacture of electrical circuits in high temperatures. One disadvantage that seems to be common to all BMI resins is the very long time of cure. The desired properties at high temperatures are achieved only by using a high temperature post-cure. The high temperature cure and post-cure cycles lead to somewhat brittle polymers with a significant amount of residual stresses. The high temperatures of curing, accompanied by a significant shrinkage, often lead to the formation of microcracks. To minimize the problems of residual stress, slow processing rates especially during the cooling phase are used.

5. Effects of manufacturing on final product

It is well known that the manufacturing process influences the quality and therefore the performance of the product. For instance, a good surface finish quality plays an important role in the mechanism of composite degradation upon exposure to the operative environment. The surface finish prevents the penetration of elements, such as dust, that produce and enhance micro-cracks within the structure. As consequence, a loss of mechanical properties occurs. Figure 4 shows the stages of the process that influence the behavior of the material and summarizes the relationship among process, material behavior and final performance. The compaction and impregnation phases govern the imperfections due to voids and dry spots. Following sections of this paragraph describe the compaction and impregnation phases.

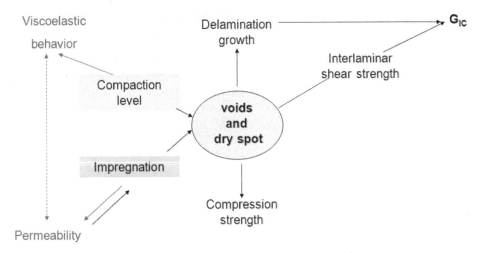

Figure 4. Relations between manufacturing phases and material behavior.

5.1. Compaction phase

The compaction phase is an important step that occurs in any production process for the manufacture of advanced composite structures. This phase usually takes place through the application of an external pressure, which produces a new arrangement of the fibers and changes of microstructure. In RTM, the compaction occurs when the mold is closed and the preform reaches the expected fiber volume fraction. Compaction produces large deformations and nonlinear viscoelastic effects in the preform. These effects are accompanied by a change in energy within the material. Some authors have modeled the phenomenon, introducing the function of free energy, but with limited success, because this hypothesis does not take into account permanent deformations of the preform. For instance, in case of fabric textile, when a multilayer preform is compacted, the fibers tend to be squashed with other fibers, producing interlayer packing or nesting phenomena depending on the deformation mode and the fibers architecture. If the compression load is removed, fibers tend to not return to their original position inside the preform.

Nesting is a phenomenon that produces irreversible mechanical and geometric changes of the preform modifying the permeability and mechanical properties of the composite material. In particular, the nesting influences the rigidity of the piece and even more the resistance. In particular, the stiffness depends on the positioning of the layers. If the layer is positioned out of phase without reducing the thickness (interlayer packing), there is no change of the mechanical properties. On the other hand, if two layers are in phase, i.e. in case of existing nesting, the difference of stiffness between the two configurations is reduced approximately of 10-20%. Nesting creates the so-called bridging between layers preventing delamination and increasing the resistance considerably, up to 10% for the layers out of phase. Moreover, the compaction induces residual stresses within the material due to the viscoelastic behavior of the fibers. The stress concentrations, also localized, can act as nucleation of cracks after the curing phase. However, during the impregnation phase a release of stress, probably due to the balance, occurs.

From a manufacturing point of view, it is evident that compaction changes the spatial arrangement of the fiber bundles, modifying substantially the morphology of the porosity. These variations alter the permeability of the preform and so the impregnation phase. Furthermore, compaction affects the adhesion between the layers, the material failure modes and the magnitude interlaminar shear stress more than the impregnation.

5.2. Impregnation phase

In RTM applications, the defects induced by the resin flow such as voids and dry spots are known as the biggest source of problems for production quality and reproducibility. The damage of the matrix is essentially related to the presence of porosity. The formation of micro-voids among fibers and dry spots are potential starting points for the propagation of cracks and delamination.

The resin flow front faces two resistance levels through the preform: the resistance between the fiber bundles and the resistance inside the fiber bundles. This means that the preforms are characterized by two different permeabilities: the permeability between the fiber bundles and the permeability inside the fiber bundles. These can be considered as a resin flow in macro-scale and micro-scale. The type of flow scale determines the potential void formation.

In the macro-scale, the formation of macro-voids can be formed when the air displaced from the resin remains trapped, or when pressure is insufficient to overcome the resistance of the preform or the viscosity is too high. In micro-scale, the porosity is given by micro-voids. Figure 5 shows the mechanism of void formation in the fiber bundle: the flow front runs around the tows and continues to impregnate the tow even after the passage of the front. If the air is not evacuated at the beginning of the filling process, it remains trapped in the tow and micro-voids occur at the center. Micro-voids formation can be also due to the macroscopic pressure drop. The pressure drop produces an apparent change in the permeability along the preform. The pressure drop can be explained by assuming the sink effect: the fiber bundles act as fluid sinks. The concept is based on the dual scale porous media. The individual fibers of the bundle are separated by a distance much smaller than that existing between two bundles. Consequently, the resin flows more easily between the fiber bundles rather than inside one of them. Therefore the liquid continues to impregnate the bundle even when the front part of the flow has passed. This means that a part of the injected fluid penetrates in a package of fiber, rather than push forward the front face of the flow. The result is that the pressure profile is influenced by the relative flow rate of the resin inside a bundle relative to the flow between bundles.

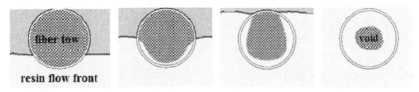

Figure 5. Mechanism of void formation in the intra-bundle.

6. Modeling of RTM process

In the RTM process, the resin flow is usually modeled with the assumption of isothermal flow, ignoring the exothermic nature of the thermosetting resin. Basing on this hypothesis, the resin flow is described by the conservation equation of mass and momentum with the boundary conditions. In order to complete the equation system, the constitutive equations of the fluid must be considered. The following section concerns the continuity and momentum equations of a resin flow within a region consisting of fibers and porosity (Figure 6).

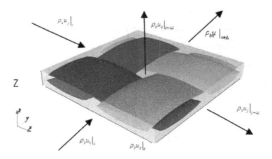

Figure 6. Control volume of porous media.

6.1. Flow model

The fiber reinforcement is a dual-scale porous material and to apply the conservation of momentum is an unrealistic approach. In fact, determining the pressure profile during the impregnation means to apply the Navier-Stocks equation for each channel of fibers network and solve the equation systems in their surroundings. This would allow to know the flow rate and the pressure within each channel. Anyway, a simple piece of preform may have millions of channels. This approach is impractical in a real case. Furthermore we have no interest in knowing the exact value of the pressure during the manufacturing process, but we need to know the relationship between pressure drop and flow rate during the progress of the flow front in the preform at a macroscopic level. For these reasons, the resin flow front is modeled by the Darcy law, considering that Reynolds number relative to the pore size is $\leq$ 1. Darcy equation (Eq. 1) describes the relationship between the flow and the pressure gradient that drives the flow through the porous medium, using the permeability parameter. This parameter, as discussed after, characterizes the fluidity of the resin through the average porosity of the fibers.

$$u = -\frac{K \cdot \nabla P}{\mu} \tag{1}$$

where u is the speed mediated on the volume, μ is the viscosity of the fluid, ∇P is the pressure gradient and K is the permeability tensor of the preform (according to the fact that fiber arrangement generally is anisotropic in porous medium).

Rewriting Eq. 1 in expanded form:

$$\begin{pmatrix} u_x \\ u_y \\ u_z \end{pmatrix} = -\frac{1}{\mu} \begin{pmatrix} K_{xx} & K_{xy} & K_{xz} \\ K_{yx} & K_{yy} & K_{yz} \\ K_{zx} & K_{zy} & K_{zz} \end{pmatrix} \begin{pmatrix} \frac{\partial P}{\partial x} \\ \frac{\partial P}{\partial y} \\ \frac{\partial P}{\partial z} \end{pmatrix} \tag{2}$$

In most RTM applications, the flow can be approximated in two-dimensional domain, since the dimensions in the plane are two or three orders of magnitude larger than the thickness. The approximation of a 3D geometry with a 2D one gives significant saves in terms of computational time and reduces the number of material parameters that need to be measured or calculated. This assumption can be considered valid since the values of permeability in the plane of the preform (in the plane in which the layers are stacked to form the thickness) are approximately of the same order of magnitude. Boundary conditions are imposed on three segments:

- injection gates
- free surface of the resin flow (i.e. on the flow front)
- walls of the mold.

The resin is injected into the mold at constant pressure or at constant flow rate, for which the gate conditions are

$$P = constant = P_{inj} \tag{3}$$

or

$$A\left(-\frac{K \cdot \nabla P}{\mu}\right) \cdot n = -Q_{inj} = constant \tag{4}$$

Some injection systems allow to vary the values of pressure or flow rate during the same injection. In such case, P_{inj} and Q_{inj} must be considered as a function of the time. This change does not alter the technique of solution, because the problem is solved as a stationary one for each time.

The pressure at the resin front flow P_{ff} can assume the value of the atmospheric and vacuum depends on the type of process, generally:

$$P_{ff}=P_{vent} \tag{5}$$

Inserts or multiple injection points can form several different resin flow fronts. When these fronts meet and they are not connected to a vent, the air moved by them remains trapped in the regions surrounded by the resin flow fronts. In this case, the boundary conditions on the front can be modified to consider that the pressure increases on the front due to the void pressure. The trapped air can be considered as an ideal gas for which the following relation applies:

$$\frac{P_{void}V_{void}}{T_{void}} = constant \tag{6}$$

However, this would require knowing a plot of the vacuum created during the impregnation of the fronts and calculate the volume at each instant of filling. In the real case, a part of this air is dissolved in the resin, reducing the pressure of the vacuum. In order to take into account this phenomenon, the system may include an equation of the resin diffusing at the mass-to-air interface and the diffusion coefficient has to be determined. But this is not a significant factor and researchers ignore this aspect and assume that the air will dissolve in the resin.

The boundary condition on the wall of the mold is the no-slip condition:

$$u_n = -\frac{1}{\mu}\left(K_{nn}\frac{\partial P}{\partial n} + K_{nt}\frac{\partial P}{\partial t}\right) = 0 \tag{7}$$

where n and t are the directions normal and tangential to the mold wall.

The continuity equation of resin flow through fibrous arrangement can be determined assuming that:

- fibers are incompressible
- variations of resin velocity and stress-strain are small in the volume control
- process is quasi-steady
- body forces, such as weight, are negligible.

The derivation is quite similar to that for any fluid within a region, except for the expression of the resin density ρ_b that is the mass of the resin divided the control volume containing both resin and fiber. In this case, u is the interstitial resin velocity within the composite. Within a fibrous region partially filled is $\rho_b=\varepsilon S\rho$ where ε is the porosity of the control volume available to be occupied by the resin, ε = (pore volume/volume control), which can be expressed as ε = 1-V_f , where V_f is the volume fraction of the fiber. S is the fraction of saturation of a porous space filled by the resin, and ρ is the true density of the resin (mass of the resin/volume of resin). The pore volume within the volume control is completely empty when $S = 0$, and completely full when $S = 1$.

The mass balance for the volume control can be expressed as:

$$\text{(net flow of mass)} = \text{(inflow)} - \text{(outflow)} - \text{(losses)} \tag{8}$$

The term of the losses is due to an internal volume V that absorbs fluid mass in a quantity equal to (x, y, z, t) per unit volume and unit time (i.e. the fiber volume). Dividing both sides by Δx, Δy and Δz and then doing the limit for Δx, Δy and $\Delta z \rightarrow 0$, the mass balance becomes:

$$\frac{\partial(\varepsilon S \rho)}{\partial t} + U \cdot \nabla(\varepsilon S \rho) + (\varepsilon S \rho)\nabla \cdot U = s \tag{9}$$

It is noted that even if the resin density ρ is constant, the porosity or saturation of the fibrous medium may not be constant within the whole region. Therefore, it cannot be zero everywhere.

7. Experimental determination of permeability

The permeability K is in general a symmetric tensor, which for an isotropic material, as random mat, is a scalar number. For a given stationary porous medium, it is necessary to know 6 scalar values K_{ji} to completely determine the tensor K. If the selected directions of the reference system are along the principal directions of the preform, the matrix becomes diagonal. Therefore, choosing the coordinate system along the main axes of the preform, the principal values of the permeability can be measured. Then, not diagonal terms can be calculated using a coordinate transformation system.

$$K = \begin{pmatrix} K_{xx} & K_{xy} & K_{xz} \\ K_{yx} & K_{yy} & K_{yz} \\ K_{zx} & K_{zy} & K_{zz} \end{pmatrix} \tag{10}$$

Permeability must be determined experimentally. There are basically two methods for this: radial flow and linear flow methods. The simplest way to determine the permeability is the use of 1D version of Darcy's equation (linear method). For a 1D flow in the direction of the axis (assumed to be the x-axis) Darcy's equation can be written as

$$\frac{dx}{dt} = -\frac{K_x}{\Phi \mu}\frac{\partial P}{\partial x} \tag{11}$$

Where $\Phi = 1 - V_f$ is the porosity of the material and V_f is the fiber volume fraction, K_x is the permeability in the x direction, μ is the fluid viscosity and $\partial P/\partial x$ is the pressure gradient between the injection point and the flow front. Since the injection pressure and the vent pressure are constant, we can transform the partial derivative in a finite difference, so we have:

$$\frac{dx}{dt} = -\frac{K_x}{\Phi\mu}\frac{\Delta P}{x} \tag{12}$$

Integrating between the position of the injection point ($x = 0$) at time $t = 0$ and the position of the flow front ($x = x$) at time $t = t$, the expression that allows to determine the permeability along a main direction of the preform is:

$$K_x = \frac{\mu\Phi slope\, of\left[x^2{}_f(t_f)\right]}{2\Delta P} \tag{13}$$

The slope of the line is calculated by plotting the position of the flow front at different times; the pressure gradient is equal to the injection pressure. In the case of orthotropic laminates, the permeability has different values in the two principal directions, and then two experiments sets must be performed.

For typical composites, the thickness is on average 2 – 3 mm. As a consequence, the transverse permeability Kz is considered negligible. This means to assume that the resin flow advances uniformly on the two external surfaces (top and bottom) and that flow velocity is an average through the thickness. However, as the thickness increases, the gradient along z (i.e. the thickness) becomes greater, due to the fact that the resin flows quickly on the upper surface, while it continues to impregnate the lower levels. Advani et al. have developed a theoretical model to calculate an average effective permeability for flat rectangular preform with thickness H formed by n layers.

$$\overline{K} = \frac{\overline{l}}{H}\sum_{j=1}^{n}\frac{K_j h_j}{l_j} \tag{14}$$

$$\overline{l} = \frac{1}{H}\sum_{j=1}^{n}\frac{l_j h_j}{n} \tag{15}$$

where h_j is the thickness of the single layer, l_j is the length of the impregnation of the single layer in the injection direction, $\overline{l}$ is the average length of impregnation in the entire laminate, K_j the permeability of the single layer.

8. Numerical analysis approach

The advantage of the numerical simulations is to help the process engineer to understand the behavior of the resin inside the mold, especially when the part geometry is complex and presents variations in permeability and fiber volume fraction. This knowledge improves the design tools and optimization of the injection scheme. General method is to use a simplified model of the resin flow reducing the problem from 3-D to 2-D and using a finite element approach and control volume (FE/CV) that does not require a re-meshing at each step. The geometry is discretized as a thin shell using triangular elements or rectangular ones. The material properties such as thickness, permeability and volume fraction of fibers can be assigned individually to each element, such as the non-uniform material properties and thickness. A linear pressure profile is assumed among the nodes of an element:

$$P^e = \sum_{i=1}^{n} N_i P_i \tag{16}$$

where n is the number of nodes of the element, P^e is the pressure inside the element, and P_i is the nodal pressure that is unknown and a function of interpolation. The Galerkin finite element method can be used to convert the governing equations to the partial derivatives in a system of algebraic equations that minimizes the error at the nodes. It can be expressed as a linear system of equations

$$[[S^e]][P] = [f] \tag{17}$$

where S is the stiffness matrix whose components are:

$$S_{ij}^e = \int_{\Omega} \left(\frac{K_{xx}}{\mu} \frac{\partial N_j}{\partial x} \frac{\partial N_i}{\partial x} + \frac{K_{yy}}{\mu} \frac{\partial N_j}{\partial y} \frac{\partial N_i}{\partial y} + \frac{K_{xy}}{\mu} \frac{\partial N_j}{\partial x} \frac{\partial N_i}{\partial y} + \frac{K_{xy}}{\mu} \frac{\partial N_j}{\partial y} \frac{\partial N_i}{\partial x} \right) d\Omega \tag{18}$$

This set of equations can be solved in every moment during the filling process. The i-th vector component is the amount of mass generated per unit time at the i-th node. The pressures are calculated at nodes by solving the set of algebraic equations described by the system.

The next step is the determination of the resin flow front progress. The approach by the control volume consists of dividing the geometry under consideration into control volumes first, and then associating one of them to each node. The flow between two control volumes is calculated by multiplying the average speed for the area connected between two control volumes. For example, the equation of the flow associated with the node i and the one associated to node j is

$$q_{ij} = - \int_{s_{ij}} \left(\frac{h}{\mu} \boldsymbol{n} \cdot [\boldsymbol{K}] \cdot \nabla P \right) ds \tag{19}$$

where s_{ij} is the boundary between two control volumes, h is the thickness of the mold cavity at the contour and n is the normal to the contour of the plane lying of the preform.

The nodal fill factor is used to track the movement of the flow front. This factor is associated with each node and represents the fraction of the volume occupied by the control fluid. The pressures are therefore determined only at the filled nodes and empty nodes are ignored. The nodes are considered partially filled close to the flow front. The progress of the flow front is considered at any instant of time t, updating the fill factor of the control volume by means of the flow between the connected nodes. Consequently, the amount of the masses is strictly taken into account. This technique can simulate the resin flow in thin cavities with high geometric complexity in 3-D. Usually the simulation process is assumed to be isothermal, because of the disadvantages in conducting non-isothermal numerical simulations. First of all, the computational efficiency and speed of convergence decrease drastically when the equations that govern the flow are solved simultaneously with the energy equation, especially with the increasing of the number of nodes. Research in the simplification of non-isothermal analysis reduced the time, but it still requires a CPU time of two or three orders of magnitude higher than the resolution of isothermal problems. Moreover, to obtain the results non-isothermal analysis requires a large number of material parameters, such as dispersion coefficient, thermal conductivity and kinetics of resin cure. For high temperature resins, such as in case of aerospace application, a good compromise would be to fill the mold under isothermal conditions, and at least considering the kinetics of cure. From a numerical point of view, this solution is very efficient because diffusive and convective problems are not faced.

9. Aerospace applications

Developing of a new composite product requires a synergy among different disciplines and sometimes entities, in other terms a Concurrent Engineering (CE) approach. This section shows some aerospace components manufactured by RTM process in CE method. The prototypes presented in the following sections are the results of years of collaboration between the School of Aerospace Engineering (currently DIAEE) of Sapienza University of Rome and Italian aerospace industries (AgustaWestland and Aermacchi).

9.1. Concurrent engineering approach

The need of high quality production, due to market pressure and high-tech demands of the aerospace field, drives to the faster and most effective design of the product. To this aim, engineers adopt new methods based on concurrent approaches to optimize the part before manufacturing the final piece. Especially from 1990s, with the explosion of the composite materials market in several industrial fields, involved people from industry and universities are working to determine the best way to apply the concurrent engineering as a systematic approach for product development. Some works are addressed to improve the management team and communication among members; others go toward building a virtual environment based on CAD systems.

The software systems for automation design tools are powerful. They can help any company to develop products of superior quality, faster and at lower cost. This implies clear advantages in terms of competitiveness. The empirical methods "test and fix" still dominate the field of molding technologies for composite materials, but this approach produces deleterious effects in terms of cost and time. Simulation studies lead to the complete optimization of the product before creating the prototype. The objective is to establish the optimal characteristics of the preform and tooling, optimizing the process parameters.

The targets that can be obtained with the use of a code are:

- reduction of design time
- reduction of the costs and consumption of materials
- optimization of the design parameters in relation to the process
- reduction of the times of modification and tuning of the molds
- integration between production process and optimization of the component with respect to constraint and loading given by specification.

About the entities involved in a project, concurrent-development teams typically exhibit the following characteristics:

- they include no more ten members
- members choose to serve on the team
- members serve from the beginning to the end of the project
- members participate in the team fulltime
- members report solely to the team leader and the leader reports to general management
- key functions - at least marketing, engineering, and manufacturing - are included in the team
- members are co-located within conversational distance of each other

The plan should give specific information about the qualification programs regarding all the entities involved in the RTM process: materials, process, tooling, tests.

Figure 7 shows the flowchart related to the CE approach set by the DIAEE of Sapienza University of Rome during the development of several RTM helicopter components. The flowchart highlights the structural and manufacturing couplings. Process parameters have been considered at the early stage of the design as a discrimination of the material selection. The resin flow behavior and the filling time of the mold were the process criteria. These parameters are directly dependent on the preform characteristics such as the textile structure, the laminate orientation and the fiber volume fraction. The preform characteristics are the first features that determine the mechanical performance of the component. In this sense, the optimized project is a compromise between structural and process optimizations and requires a concurrent engineering design and manufacturing.

9.2. Strake

The strake is a typical aerodynamic surface used in both supersonic and subsonic vehicles to improve the stability and/or reduce the drag. The thickness of the structure is around 2-3 mm and usually made in aluminum alloy. In this case, the selected resin was a mono-component benzoxazine polymer (Henkel Epsilon 99110). The recommended process conditions were typical infusion temperature at 90 °C and curing at 180 °C for 90 min. The fiber reinforcement used was a carbon textile HEXCEL G0926 5H Satin. The preforms were realized manually overlaying several layers adopting a predefined lamination sequences. The lamination sequences were determined by structural analysis. The injection equipment was a commercial Hyperject machine for mono-component polymers. The Hyperject injects the resin at constant pressure and it is provided with heated dispending resin. The resin was loaded, degassed and heated inside the Hyperject. When both the temperatures of the resin and the mold were reached, the resin was pumped inside the mold until the resin came out from the vent. Molds were studied to be multi-components in order to make easily the demolding.

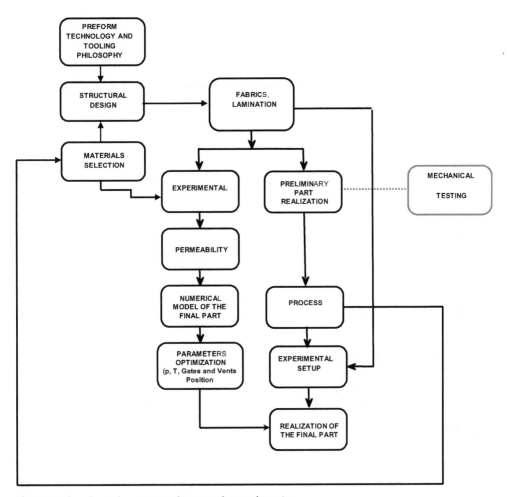

Figure 7. Flowchart of concurrent design and manufacturing

The numerical analysis was performed by commercial codes based on a finite element analysis-control volume (FEM-CV). Modelling of the flow was to allow investigating the resin impregnation process and strategically designing gates and vents and injection scheme in order to optimally fill the composite part without any dry spot. The analysis took into account the most relevant process parameters such as injection pressure and material features like the resin viscosity and permeability of the preform. The FEM-CV analysis was isothermal and provided detailed information about the pressure field, flow front patterns, and strategic injection scheme. The different choice of gate positions affects very much the filling time and the quality of the finished part. For these reasons, different injection schemes were considered. All simulations were performed using the resin viscosity value at the mold temperature.

Figure 8 shows the simulation result of the best injection scheme. In particular, the simulation represents the trend of the filling time. Figure 9 shows the prototype of the strake after demolding.

Figure 8. Injection scheme of the mold as determined by numerical simulation. On the top: filling time trend through the preform. On the bottom at left: carbon fiber reinforcement before closing mold.

Figure 9. Strake of Aermacchi nacelles after demolding.

9.3. Inboard flaperon

The aim of the work was to realize a primary structure of a helicopter by RTM process in order to compare its properties with those of the parts made with classical hand lay-up technique. The chosen component was the inboard flaperon of the BA609 (Bell Agusta), which is an aerodynamic control surface that presents different critical features. In fact, the inboard flaperon is a primary component so the reliability must be nearly absolute. Further, it is a large part and involved a complicated lamination sequence that can affect the manufacturing process. It is evident in Figure 10 the presence of three different lamination sequences that correspond to different permeability, which was determined experimentally.

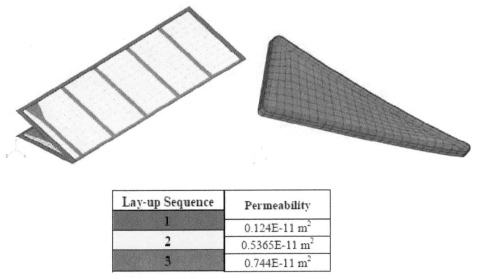

Lay-up Sequence	Permeability
1	0.124E-11 m^2
2	0.5365E-11 m^2
3	0.744E-11 m^2

Figure 10. Finite element model of the inboard flaperon with a particular on a rib (red component on the right). The different colors individuate the different permeability values associated to the lamination sequence.

In the first scheme, the injection points were placed on the two leading edge borders of the skin in correspondence of the rib reinforces and the venting points on the trailing edge line. As shown in Figure 11, the total filling time given by this configuration is very long (more than 1.5 h) and comparable with the resin pot life limit.

A second scheme has been analyzed. In this case five injection points were placed in the middle of each skin side and between the rib reinforces. The venting points were located both on the trailing and leading edge lines. With this second scheme the filling time was considerably reduced down to 20 min in good agreement with the resin processability. From the results of the previous simulation it was noted that much time was needed to fill a very small area near the outbord side. For this reason a further optimization was carried out including a sixth injection point in this area. With this simple modification the total filling resulted to be less than half of the previous simulation (Figure 12).

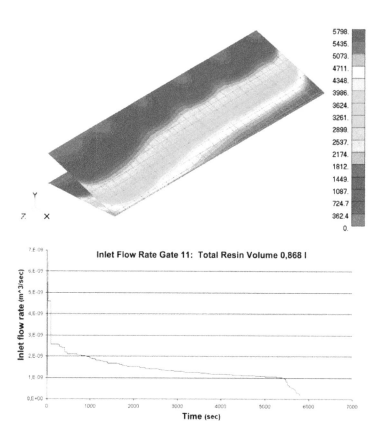

Figure 11. Filling time with the first injection scheme and predicted resin flow rate during the injection with the first scheme.

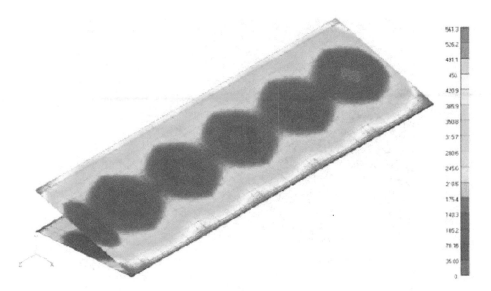

Figure 12. Filling time trend for the injection scheme selected for the mold.

The dry fabric plies for the construction of the preform were cut with the GFM ultrasonic cutting machine using a multilayer method. In this way some areas of the preform were cut directly in the final lamination sequence allowing for considerable time saving in the preform assembly phase. The different plies and sequences were sticked together on a flat surface using an adequate agent (binder). It was used the same cut program and operative cycle of the original part. In Figure 13, the glass layers used to replace the syntactic foam can be seen. The use of the glass mat strongly improves the permeability of the preform allowing very much shorter injection times. Further, the syntactic foam was incompatible with the RTM process.

Figure 13. Particular of preform preparation. On the right: glass layers used to replace the syntactic foam of the original component.

Figure 14. On the top: moment of the impregnation phase. On the bottom: the final component after demolding.

9.4. Bracket

The bracket is a secondary element that is present in large numbers in a helicopter. Usually it is made by metal alloy. The aim to re-design the bracket in composite materials is to reduce the weight and time of production. In this case, a consistent saving can be reached adopting an automatic textile process, which avoids the long time due to the lamination. The stitching was the selected preform technique. Figure 15 shows the cut of the layers and the preform after the assembling. Figure 16 shows the realization of the component during the impregnation phase and the part demoded after the curing step. In this case, the use of RTM process permits to save 30% weight with respect to its equivalent in metal alloy and reduces drastically the production cycle.

Figure 15. Realization of the preform: cut and stitching.

Figure 16. Realization of the bracket by RTM process: resin injection (left), bracket demolded (right).

Author details

Susanna Laurenzi and Mario Marchetti

Department of Astronautic Electrical and Energy Engineering, Sapienza Università di Roma, Italy

10. References

Acheson JA, Simacek P, Advani SG. The implications of fiber compaction and saturation on fully coupled VARTM simulation. Composites Part A: Applied Science and Manufacturing. 2004;35(2):159-169.

Amico S, Lekakou C. An experimental study of the permeability and capillary pressure in resin-transfer moulding. Composites Science and Technology. 2001;61(13):1945-1959.

Baker AA, Callus PJ, Georgiadis S, Falzon PJ, Dutton SE, Leong KH. An affordable methodology for replacing metallic aircraft panels with advanced composites. Composites Part A: Applied Science and Manufacturing. 2002;33(5):687-696.

Barlow D, Howe C, Clayton G, Brouwer S. Preliminary study on cost optimisation of aircraft composite structures applicable to liquid moulding technologies. Composite Structures. 2002;57(1–4):53-57.

Belov EB, Lomov SV, Verpoest I, Peters T, Roose D, Parnas RS, et al. Modelling of permeability of textile reinforcements: lattice Boltzmann method. Composites Science and Technology. 2004;64(7–8):1069-1080.

Bickerton S, Abdullah MZ. Modeling and evaluation of the filling stage of injection/compression moulding. Composites Science and Technology. 2003;63(10):1359-1375.

Bickerton S, Buntain MJ, Somashekar AA. The viscoelastic compression behavior of liquid composite molding preforms. Composites Part A: Applied Science and Manufacturing. 2003;34(5):431-444.

Bickerton S, Buntain MJ. Modeling forces generated within rigid liquid composite molding tools. Part B: Numerical analysis. Composites Part A: Applied Science and Manufacturing. 2007;38(7):1742-1754.

Brouwer WD, van Herpt ECFC, Labordus M. Vacuum injection moulding for large structural applications. Composites Part A: Applied Science and Manufacturing. 2003;34(6):551-558.

Buntain MJ, Bickerton S. Modeling forces generated within rigid liquid composite molding tools. Part A: Experimental study. Composites Part A: Applied Science and Manufacturing. 2007;38(7):1729-1741.

Cairns DS, Humbert DR, Mandell JF. Modeling of resin transfer molding of composite materials with oriented unidirectional plies. Composites Part A: Applied Science and Manufacturing. 1999;30(3):375-383.

Calado VnMA, Advani SG. Effective average permeability of multi-layer preforms in resin transfer molding. Composites Science and Technology. 1996;56(5):519-531.

Chen B, Chou T-W. Compaction of woven-fabric preforms: nesting and multi-layer deformation. Composites Science and Technology. 2000;60(12–13):2223-2231.

Chen B, Lang EJ, Chou T-W. Experimental and theoretical studies of fabric compaction behavior in resin transfer molding. Materials Science and Engineering: A. 2001;317(1–2):188-196.

Chen Z-R, Ye L, Kruckenberg T. A micromechanical compaction model for woven fabric preforms. Part I: Single layer. Composites Science and Technology. 2006;66(16):3254-3262.

Chen Z-R, Ye L. A micromechanical compaction model for woven fabric preforms. Part II: Multilayer. Composites Science and Technology. 2006;66(16):3263-3272.

Correia NC, Robitaille F, Long AC, Rudd CD, Šimáček P, Advani SG. Analysis of the vacuum infusion moulding process: I. Analytical formulation. Composites Part A: Applied Science and Manufacturing. 2005;36(12):1645-1656.

Dobyns A, Rousseau CQ, Minguet P. 6.12 - Helicopter Applications and Design. In: Editors-in-Chief: Anthony K, Carl Z, editors. Comprehensive Composite Materials, Oxford: Pergamon; 2000. p. 223-242.

Gutowski TG, Cai Z, Kingery J, Wineman SJ. Resin flow/fiber deformation experiments. Journal Name: SAMPE Q; (United States); Journal Volume: 17:4. 1986:Medium: X; Size: Pages: 54-58.

Haider M, Hubert P, Lessard L. An experimental investigation of class A surface finish of composites made by the resin transfer molding process. Composites Science and Technology. 2007;67(15–16):3176-3186.

J.M. Lawrence PS, S. Laurenzi, S. G. Advani. Flow modelling of the compression resin transfer molding process. The 8th International Conference on Flow Processes in Composite Materials, Douai, France2006.

Kruckenbrg T. Paton R. Resin Transfer Moulding for Aerospace Structures: Kluwer Academic Publishers.

Laskoski M, Dominguez DD, Keller TM. Synthesis and properties of a liquid oligomeric cyanate ester resin. Polymer. 2006;47(11):3727-3733.

Laurenzi S, Di Nallo D, Marchetti M, Lalia Morra E, Anamateros E. Manufacturing approach to realize a prototype of a helicopter transmission component by composite materials. 31st European Rotorcraft Forum, Florence, Italy2005.

Laurenzi S, Di Nallo D, Valente F, Marchetti M, Lalia Morra E, Anamateros E. Re-design of a helicopter component transmission by composite material. 45st Israel Annual Conference on Aerospace Sciences, Haifa, Israel2005.

Laurenzi S, Griccini M, Lalia Morra E, Anamateros E, Marchetti M. Processability analysis of thick braided composites manufactured with rtm technology. 7th International Conference on Flow Processes in Composite Materials, Newark, Delaware, USA2004.

Laurenzi S, Marchetti M, Anamateros E. Liquid Composite Molding for aeronautical application. International Council of the Aeronautical Sciences Hamburg, Germany2006.

Laurenzi S. Liquid composite molding for aerospace applications. Case studied:Helicopter A 109 gearbox. Sapienza University of Rome, School of Aerospace Engineering, 2007.

Lomov SV, Verpoest I, Peeters T, Roose D, Zako M. Nesting in textile laminates: geometrical modelling of the laminate. Composites Science and Technology. 2003;63(7):993-1007.

Luo Y, Verpoest I, Hoes K, Vanheule M, Sol H, Cardon A. Permeability measurement of textile reinforcements with several test fluids. Composites Part A: Applied Science and Manufacturing. 2001;32(10):1497-1504.

Marchetti M, Cutolo D. Tecnologie dei materiali composite: ESA Grafica; 1991.

Merotte J, Simacek P, Advani SG. Resin flow analysis with fiber preform deformation in through thickness direction during Compression Resin Transfer Molding. Composites Part A: Applied Science and Manufacturing. 2010;41(7):881-887.

Mouritz AP, Bannister MK, Falzon PJ, Leong KH. Review of applications for advanced three-dimensional fibre textile composites. Composites Part A: Applied Science and Manufacturing. 1999;30(12):1445-1461.

Ngo ND, Tamma KK. Microscale permeability predictions of porous fibrous media. International Journal of Heat and Mass Transfer. 2001;44(16):3135-3145.

Noor AK, Venneri SL, Paul DB, Hopkins MA. Structures technology for future aerospace systems. Computers & Structures. 2000;74(5):507-519.

Park CH, Lee WI, Han WS, Vautrin A. Simultaneous optimization of composite structures considering mechanical performance and manufacturing cost. Composite Structures. 2004;65(1):117-127.

Parseval YD, Pillai KM, Advani SG. A Simple Model for the Variation of Permeability due to Partial Saturation in Dual Scale Porous Media. Transport in Porous Media. 1997;27(3):243-264.

Potter KD. The early history of the resin transfer moulding process for aerospace applications. Composites Part A: Applied Science and Manufacturing. 1999;30(5):619-621.

Reduced cost, higher performance RTM. Reinforced Plastics. 1997;41(9):48-54.

Reia da Costa EF, Skordos AA. Modelling flow and filtration in liquid composite moulding of nanoparticle loaded thermosets. Composites Science and Technology. 2012;72(7):799-805.

S. G. Advani EMS. Process Modeling in Composites Manufacturing: Marcel Dekker press; 2002.

S. G. Advani MVB, R. Parnas. Resin Transfer Molding. Flow and Rheology in Polymeric Composites Manufacturing, Amsterdam: Elsevier Publishers; 1994.

Saunders RA, Lekakou C, Bader MG. Compression and microstructure of fibre plain woven cloths in the processing of polymer composites. Composites Part A: Applied Science and Manufacturing. 1998;29(4):443-454.

Saunders RA, Lekakou C, Bader MG. Compression in the processing of polymer composites 1. A mechanical and microstructural study for different glass fabrics and resins. Composites Science and Technology. 1999;59(7):983-993.

Simacek P, Advani SG. A numerical model to predict fiber tow saturation during liquid composite molding. Composites Science and Technology. 2003;63(12):1725-1736.

Šimáček P, Advani SG. Desirable features in mold filling simulations for Liquid Composite Molding processes. Polymer Composites. 2004;25(4):355-367.

Slade J, Pillai KM, Advani SG. Investigation of unsaturated flow in woven, braided and stitched fiber mats during mold-filling in resin transfer molding. Polymer Composites. 2001;22(4):491-505.

Smith P, Rudd CD, Long AC. The effect of shear deformation on the processing and mechanical properties of aligned reinforcements. Composites Science and Technology. 1997;57(3):327-344.

Soutis C. Carbon fiber reinforced plastics in aircraft construction. Materials Science and Engineering: A. 2005;412(1–2):171-176.

Tan H, Roy T, Pillai KM. Variations in unsaturated flow with flow direction in resin transfer molding: An experimental investigation. Composites Part A: Applied Science and Manufacturing. 2007;38(8):1872-1892.

Tari MJ, Bals A, Park J, Lin MY, Thomas Hahn H. Rapid prototyping of composite parts using resin transfer molding and laminated object manufacturing. Composites Part A: Applied Science and Manufacturing. 1998;29(5–6):651-661.

Permissions

The contributors of this book come from diverse backgrounds, making this book a truly international effort. This book will bring forth new frontiers with its revolutionizing research information and detailed analysis of the nascent developments around the world.

We would like to thank Ning Hu, Ph.D., for lending his expertise to make the book truly unique. He has played a crucial role in the development of this book. Without his invaluable contribution this book wouldn't have been possible. He has made vital efforts to compile up to date information on the varied aspects of this subject to make this book a valuable addition to the collection of many professionals and students.

This book was conceptualized with the vision of imparting up-to-date information and advanced data in this field. To ensure the same, a matchless editorial board was set up. Every individual on the board went through rigorous rounds of assessment to prove their worth. After which they invested a large part of their time researching and compiling the most relevant data for our readers. Conferences and sessions were held from time to time between the editorial board and the contributing authors to present the data in the most comprehensible form. The editorial team has worked tirelessly to provide valuable and valid information to help people across the globe.

Every chapter published in this book has been scrutinized by our experts. Their significance has been extensively debated. The topics covered herein carry significant findings which will fuel the growth of the discipline. They may even be implemented as practical applications or may be referred to as a beginning point for another development. Chapters in this book were first published by InTech; hereby published with permission under the Creative Commons Attribution License or equivalent.

The editorial board has been involved in producing this book since its inception. They have spent rigorous hours researching and exploring the diverse topics which have resulted in the successful publishing of this book. They have passed on their knowledge of decades through this book. To expedite this challenging task, the publisher supported the team at every step. A small team of assistant editors was also appointed to further simplify the editing procedure and attain best results for the readers.

Our editorial team has been hand-picked from every corner of the world. Their multi-ethnicity adds dynamic inputs to the discussions which result in innovative outcomes. These outcomes are then further discussed with the researchers and contributors who give their valuable feedback and opinion regarding the same. The feedback is then collaborated with the researches and they are edited in a comprehensive manner to aid the understanding of the subject.

Apart from the editorial board, the designing team has also invested a significant amount of their time in understanding the subject and creating the most relevant covers. They scrutinized every image to scout for the most suitable representation of the subject and create an appropriate cover for the book.

The publishing team has been involved in this book since its early stages. They were actively engaged in every process, be it collecting the data, connecting with the contributors or procuring relevant information. The team has been an ardent support to the editorial, designing and production team. Their endless efforts to recruit the best for this project, has resulted in the accomplishment of this book. They are a veteran in the field of academics and their pool of knowledge is as vast as their experience in printing. Their expertise and guidance has proved useful at every step. Their uncompromising quality standards have made this book an exceptional effort. Their encouragement from time to time has been an inspiration for everyone.

The publisher and the editorial board hope that this book will prove to be a valuable piece of knowledge for researchers, students, practitioners and scholars across the globe.

List of Contributors

Mingchao Wang, Cheng Yan and Lin Ma
School of Chemistry, Physics and Mechanical Engineering, Science and Engineering Faculty, Queensland University of Technology, Brisbane, Australia

Sumanta Bhandary and Biplab Sanyal
Department of Physics and Astronomy, Uppsala University, Box 516, 751 20 Uppsala, Sweden

Ilya Mazov
Boreskov Institute of Catalysis, Novosibirsk, Russia
National Research Technical University "MISIS", Moscow, Russia

Vladimir Kuznetsov
Boreskov Institute of Catalysis, Novosibirsk, Russia

Anatoly Romanenko
Nikolaev Institute of Inorganic Chemistry, Novosibirsk, Russia

Valentin Suslyaev
National Research Tomsk State University, Tomsk, Russia

Marcin Molenda, Michał Świętosławski and Roman Dziembaj
Jagiellonian University, Faculty of Chemistry, Krakow, Poland

David Alejandro Arellano Escárpita, Diego Cárdenas, Hugo Elizalde and Ricardo Ramirez
Mechatronics Engineering Department, Instituto Tecnológico y de Estudios, Superiores de Monterrey, Campus Ciudad de México, Col. Ejidos de Huipulco, Tlalpan, México D.F., Mexico

Oliver Probst
Physics Department, Instituto Tecnológico y de Estudios Superiores de Monterrey, Campus Monterrey, Monterrey, N.L., Mexico

Milan Žmindák and Martin Dudinský
University of Žilina, Slovakia

Andrey Radchenko
Tomsk State University of Architecture and Building, Russia

Pavel Radchenko
Institute of Strength Physics and Materials Science of SB RAS, Russia

F. Wang
Southwest University, China

J.Q. Zhang
Shanghai University, China
Shanghai Key Laboratory of Mechanics in Energy Engineering, China

Yuan Li
Department of Nanomechanics, Tohoku University, Aramaki-Aza-Aoba, Aoba-ku, Sendai, Japan

Sen Liu
Department of Mechanical Engineering, Chiba University, Yayoi-cho, Inage-ku, Chiba, Japan

Ning Hu
Department of Mechanical Engineering, Chiba University, Yayoi-cho, Inage-ku, Chiba, Japan

Weifeng Yuan and Bin Gu
School of Manufacturing Science and Engineering, Southwest University of Science and Technology, Mianyang, P.R. China

E. Dado
Netherlands Defence Academy, Breda, The Netherlands

E.A.B. Koenders
Delft University of Technology, Delft, The Netherlands
COPPE-UFRJ, Programa de Engenharia Civil, Rio de Janeiro, Brazil

D.B.F. Carvalho
(PUC-Rio), Rio de Janeiro, Brazil

Susanna Laurenzi and Mario Marchetti
Department of Astronautic Electrical and Energy Engineering, Sapienza Università di Roma, Italy

Printed in the USA
CPSIA information can be obtained
at www.ICGtesting.com
JSHW011436221024
72173JS00004B/832